This book is dedicated to Dr Sandra Ward in recognition of her services to the British Library as chair of the Science Reference and Information Service Advisory Committee 1986–96.

EVIDENCE-BASED MEDICINE

AN OVERVIEW AND GUIDE TO THE
LITERATURE

By LESLEY GRAYSON

THE BRITISH LIBRARY

Evidence-based Medicine: an overview and guide to the literature
ISBN 0–7123–0836–9

Published by

The British Library
Science Reference and Information Service
25 Southampton Buildings
London WC2A 1AW

© 1997 The British Library Board

British Library Cataloguing-in-Publication Data
A catalogue record for this book is available from the British Library

For further information on publications produced by The British Library Science
Reference and Information Service (SRIS) contact Paul Wilson, SRIS,
25 Southampton Buildings, London WC2A 1AW. Tel: 0171–412 7472.

Desktop publishing by Tradespools Ltd, Frome, Somerset.

Printed by Hobbs the Printers Ltd, Totton, Hampshire SO40 3YS.

EVIDENCE-BASED MEDICINE

CONTENTS

ABOUT THE AUTHOR

Lesley Grayson is a freelance writer and editor specialising in literature-based guides on topics of interest for policy makers, researchers and practitioners. She is editor of The British Library journal *Science, Technology and Innovation*; editor of the public policy review series *Inlogov Informs* published by the University of Birmingham Institute of Local Government Studies; and author of a range of information guides, principally on environmental and social policy issues.

ABOUT PROFESSOR SHELDON

Trevor Sheldon is the Director of the NHS Centre for Reviews and Dissemination at the University of York. This is funded by the UK National Health Service Research and Development Programme to put together and disseminate research intelligence to NHS decision makers. He is trained in medicine, economics and medical statistics. His main research interests include evaluating the effectiveness and cost-effectiveness of health care interventions, helping to influence policy and practice to be more scientifically based, resource allocation, health care reform, performance measurement and methods of systematic reviews.

INTRODUCTION

Trevor Sheldon
Director, NHS Centre for Reviews and Dissemination

Over the last 30 years health care in general, and medicine in particular, has received critical attention which has, at times, affected professional hegemony. The role of medicine in contributing to significant improvements in the health and welfare of populations in developed industrialised countries was challenged. Not only did a more modest picture of the impact of medicine emerge, but analysis of clinical practice patterns and the behaviour of medical professionals demonstrated that the practice of medicine was often a reflection of social forces influenced by fashion, professional power and (in fee-for-service systems) income maintenance, rather than being based on sound empirical evidence of net benefit for patients or populations.

Research-based decision making

The publication of Archie Cochrane's radical critique, *Efficiency and effectiveness*, 25 years ago stimulated a penetrating examination of the degree to which medical practice is based on robust research demonstrating clinical effectiveness. The findings that even the most commonly used procedures or therapies are not necessarily those shown by research studies to be the most efficacious, and that a not insubstantial amount of practice has not been particularly well evaluated, has led to pressure for medicine to become more 'evidence-based'. Instead of practice being dominated by opinion (possibly ill-informed) and by consensus formed in poorly understood ways by 'experts', the idea is to shift the centre of gravity of health care decision making towards an explicit consideration and incorporation of research evidence.

To any member of the public this would surely seem to be a welcome and unexceptional development. However, the evidence-based movement has generated much controversy in professional and policy circles. The direct implication that professionals and experts may have got it wrong, sometimes dramatically so, is a bitter pill to swallow and inevitably produces hostile reactions. The potentially 'democratising' effect of an evidence-based approach - in which anyone with the ability to analyse the evidence, or with access to systematic reviews of it, can critically scrutinise practice - threatens to weaken the privileged positions of some of the professional elites.

Commercial organisations which produce and market drugs and equipment may also feel threatened. Not only may the claims they make for their products (often based on flimsy research and selective use of research results) be shown to be exaggerated, they may be forced into investing more heavily in high quality and longer term evaluative research as the rules of evidence become more exacting. The evidence-based approach also renders the cadres of 'captured' researchers and expert clinicians less credible and useful to industry as (often unwitting) marketing agents.

These debates, though important, are reactive and predictable. They will tend to diminish over time as the culture changes and foci of influence are reconfigured. The traditional elites may absorb, incorporate and possibly transform the new agenda, thereby preserving their influence (for example, the Medical Royal Colleges have embraced and taken over production of evidence-based clinical practice guidelines) or new centres of influence may become established (for example, the Centres for Evidence-Based Medicine and Child Health).

The dangers of politicisation

Other aspects of the evidence-based movement are more complex and challenging. Evidence-based medicine (EBM) is embraced to a varying extent, and under different banners and in different forms, by government health departments and health care funders in Europe and North America. EBM is seen by them as a way not only of promoting quality but also of containing costs through the restriction of practice to a more limited set of 'effective' procedures. This may be achieved, for example, by the development of evidence-based clinical practice guidelines or through coverage decisions (in insurance-based systems).

The potential for investment in clinical evaluative research to reduce health care expenditure is, however, largely illusory. Whilst there are several examples of savings from the implementation of research findings (such as the use of therapy to eradicate heliobacter pylori so avoiding the need for more expensive long term acid suppression treatment), the huge demands for health care and health technologies mean that this approach cannot save money unless everything else is held constant. Altering features of the organisation and financing of health care such as the introduction of global budgets (cash limits) and single pipe funding, which help control the supply of and demand for health care, are more effective at containing costs than EBM.

Research on effectiveness, often incorrectly seen (especially in the USA) as a technical fix for social and political problems with health care systems, can at best help to ensure that whatever resources we do use are spent in a way which is likely to produce the greatest positive impact on health and welfare. The use of EBM by government or health care purchasers as a tool to 'save' money means that the evidence-based project will be seen as a weapon of politicians or managers to deprive the population of health care, and the professionals of a say in the care of their patients.

No sensible proponent of the EBM approach would deny the important role of clinical judgement in the judicious use of research evidence. However, in the hands of those unaware of the subtleties or complexities of practice and the limitations of the research evidence (and the sorts of questions which research can sensibly address), summaries of research evidence can be very blunt instruments indeed. The use, or rather mis-use, of research as a fig leaf for, or as a way of legitimising, resource savings is likely to discredit the whole EBM programme. This is already being seen in the USA where managed care organisations are imposing draconian and often unreasonable limitations on what health care they will cover. Some of this may be occurring under the cover of 'evidence'. The politicisation of EBM may defeat its original aims, and may also result in those activities of health and social care which are less amenable to empirical investigation being undervalued.

What is evidence?

Several other problems accompany the EBM project. Most important is the epistemological issue of what constitutes evidence, and what constitutes sufficient evidence to justify health care practices. The epistemological mismatch between researchers and clinicians has been well described. Clinicians often lay more emphasis on their own experience drawn from individual cases, and may have a more mechanistic view of treatments, than evaluation scientists who work on the basis of average effects on large groups. Experience can be misleading as an indication of the effectiveness of a treatment especially when the effects are small, statistical and long term. However, if ignored totally, the knowledge accumulated through experience will be wasted, and practice become the poorer for it.

Previously, the opinions or experiences of a few (often self-proclaimed) experts were thought sufficient to produce guidelines to be followed by others. However, there can be little scientific justification for assuming that one's own experience is more valid than that of others, without reference to research-based knowledge. The challenge for EBM is to find the proper role for both sorts of evidence (research-based and experiential) and an explicit and scientific approach to combining them in practice. Research evidence is more likely to be useful in providing general statements of benefit and costs which can be interpreted and used locally in making practice decisions.

Research evidence is not homogeneous or monolithic. There is a range of methods and designs which have been used to evaluate treatments or preventive interventions. Within any type of design particular studies have been carried out with variable quality and attention to methodological rigour. The case for randomised controlled trials (RCTs) being regarded as the most reliable design is strong because they attempt to compare like with like and so are less susceptible to a variety of biases. However, many RCTs are of poor quality and usually ignore more qualitative aspects, particularly of outcome assessment. This is not inherent in the design but reflects the poor quality of, and methodological biases in, evaluative health care research generally.

EBM is only as strong as the research base which it wishes, justifiably, to thrust more centrally into the clinical decision making process. However, the variable quality of the research and the lack of an agreed scientific framework for appraising quality and converting it into recommendations limits our ability to operationalise the concepts of EBM. There is a temptation to avoid these complexities by number counting sorts of exercises which (say) estimate the percentage of treatments used which have been evaluated by an RCT, a rather reductionist activity that may provide useful context but fails to take us much further forward.

EBM is dependent on the nature of primary research which it attempts to uncover by means of systematic reviews of the evidence, and get more generally disseminated to practitioners. However, the evidence base is not naturally occurring, but created by researchers and by the organisations that employ them. Research is only carried out if researchers are interested in a particular question and if there are the necessary resources to fund the work. Neither of these is neutral and may reflect to some extent commercial and professional interests.

Thus much of the available research literature represents attempts by companies to get their products licensed, covered and promoted. Many of the important questions for the health care system and for patients will therefore never have been addressed because no one has been sufficiently interested to fund such research. For example, there is little research on the effects of simple paracetamol on arthritis because it is out of patent, whereas there are many trials of patented (and more expensive) non-steroidal anti-inflammatory drugs. Similar biases are seen against more health-promoting interventions. Studies may be biased against examining outcomes valued by patients but viewed as less important by clinicians. Whilst EBM attempts to paint an accurate picture of the effectiveness of health care interventions, the paint and the canvas are determined by others, many of whom have different (even competing) objectives.

A broader focus

The activity which goes under the banner of EBM is focused almost entirely upon research evaluating the clinical effectiveness of treatments, ignoring the issue of cost-effectiveness. Information on the resources used to achieve health care benefits is important because of their opportunity cost: resources used for one type of intervention are not available for an alternative use. Clinical guidelines usually ignore cost-effectiveness which may result in the sub-optimal allocation of resources. Only if EBM takes into account costs, and embraces a broader concept of the value of health care outcomes, is it likely to succeed in helping to push the health service into more globally rational and efficient decision making.

Though EBM has mainly been aimed at individual clinical decision makers, its principles can surely be applied to other areas such as health care management and policy making where, traditionally, decisions have been more politically influenced and are strongly dependent on fashions. These reflect the illusory certainties peddled by 'management gurus' based on a level of research or empirical evidence which would be risible even to an opinion-oriented clinical audience. Examples include the possible over-emphasis on the concentration of acute hospital services on the basis of an unproven belief that mergers will save money, the introduction of performance related pay and local pay bargaining, and of managed competition.

Conclusions

Many of the issues raised above are not criticisms of the EBM programme as such but represent challenges which will have to be faced as it develops into an organic part of health care decision making. In order to succeed in this process of maturation and reflection on the complexities of the task (for EBM is still in its infancy) the momentum needs to be sustained. Government in general, and the NHS R&D programme in particular, will need to maintain a commitment to funding those who seek to pull together the best evidence and develop methodologies for appraising the research and helping to translate it into recommendations for practice.

Of central importance to the process will be the ability of researchers to maintain independence from the various forces - professional, commercial and governmental - that might seek to influence the results of their work. If EBM provides one of the arenas in which battles over coverage, spending and rationing are going to be acted out, there will be a strong incentive to try and control what EBM researchers do and say. Government may be tempted to prevent publication and dissemination of publicly funded research syntheses that recommend actions which involve more spending, or which contradict existing policy, or which are otherwise inconvenient. Industry has a track record of hiding unfavourable research results and influencing the nature of research in order to cast its products in a more favourable light, while professional organisations rarely look kindly on members who go public with findings at odds with the perceived interests of the group.

EBM requires not only that the production of primary research and systematic reviews of this research be independent, relevant to public needs and of high methodological quality, but also that the results are easily available for decision makers. This means that databases must be designed to be accessible, and to provide information in useful forms for professionals and, where sensible, the public and that copies of research results should be lodged with central organisations such as the British Library. Ensuring access and enhancing bibliographic control (for example, over 'grey' or semi-published material) must become a priority over the next decade. Likewise increased information on ongoing research is an important goal to ensure identification of impending new research results. The possibility that the NHS Executive will abandon its plan to develop the NHS Net with free access to these sorts of materials would mark its failure to match the rhetoric of clinical and cost-effectiveness with investment in the infrastructure to help deliver the necessary information.

The rise of EBM also presents a major challenge to librarians in the NHS and in universities whose traditional role of information brokers now takes on even greater importance. Librarians need to ensure that they have the necessary skills and understanding to help researchers access the results of research in a systematic way and to distinguish between the quality (or provenance) of information sources as well as being able to appreciate the importance and relevance of systematic reviews and of issues such as study design and quality of research evidence. In parallel to this the NHS, which in many areas has been blind to the importance of information services, will need to invest more in high quality professional networks of librarians and libraries. This book will help information staff and decision makers in the health service understand more about EBM, the opportunities it presents, and the problems to be tackled. It is also a valuable source document for those seeking to access more detailed information.

EXECUTIVE SUMMARY

Evidence-based medicine (EBM) has been described by one of its foremost proponents as 'the conscientious, explicit and judicious use of current best evidence in making decisions about the care of individual patients', at first sight a fairly straightforward description of what most people believe doctors to have been doing since medical practice began. In reality, the implementation of evidence-based health care offers a significant challenge to the traditional culture and practice of medicine, particularly in view of its undoubted attractions to policy makers desperate to control health service expenditure and improve efficiency and effectiveness. It has provoked a sometimes ill-tempered professional debate about what constitutes good medical practice, and has become inextricably linked with the debate over priority-setting in a resource-constrained health service. Like many radical innovations, EBM seems a commonsense notion to many people until they come to face the myriad practical difficulties of changing attitudes and behaviour.

In a complex and rapidly changing field like EBM it is impossible for a literature guide such as this to do more than scratch the surface. What follows is a snapshot of developments at a particular point in time, not a detailed and comprehensive review of the relevant literature which is vast and multi-faceted. It makes no claims to have covered all the many dimensions of EBM, and there is no doubt that during the interval between delivery of copy and final publication things will have moved on yet again. However, it attempts to introduce at least some of the many issues involved in the EBM debate and identify sources of information (conventional and electronic) that one might turn to in developing a more informed understanding of this important development in health care.

- **Chapter 1** begins by looking briefly at some of the reasons why doctors may find it difficult to follow an evidence-based approach to practice, including problems of information overload. Though most practitioners do draw on evidence derived from research and from their own observations when treating patients, relatively few do so in a systematic way. As a result - according to the EBM community - their standards of professional competence decline progressively once they have completed their initial training. The evidence-based approach, pioneered by Archie Cochrane and latterly by David Sackett, was initially described as a paradigm shift in medical practice but is now more usually promoted as a complement to the traditional approach. Nonetheless, it offers a significant challenge to time-honoured authority structures within medicine, and remains a source of controversy.

- **Chapter 2** discusses how EBM as a purely professional response to the desire to improve the quality of care for individual patients has been seized upon (some say hijacked) by health service policy makers and managers in pursuit of a more efficient and effective health service. With its emphasis on eradicating useless or wasteful practices and concentrating on those which deliver the best outcome, EBM has enormous attractions as a tool for targeting scarce NHS resources. The drive for improved clinical- and cost-effectiveness is central to the health service in the 1990s but, in the process, there is a danger that EBM's original principles will be subverted. There is concern that it may be used to justify cutting expenditure on ineffective or less effective treatments, while its role in highlighting the need for investment in more effective ones may be downplayed. Perhaps most worrying is the possibility that EBM may be used as a smokescreen for rationing based on financial rather than clinical criteria.

- **Chapter 3** outlines some of the major institutional sources of evidence-based health care information including the Centre for Evidence-Based Medicine in Oxford, the NHS Centre for Reviews and Dissemination in York, and the international Cochrane Collaboration. It makes no attempt at comprehensive coverage of what is a confusing, fragmented and constantly evolving picture, and perhaps the best way of keeping up to date with what is going on in EBM is to use one of the gateway services on the Internet. The OMNI (Organising Medical Networked Information) Project at the National Institute for Medical Research, and the service produced by Andrew Booth of the Sheffield Centre for Health and Related Research are the most useful of these, and access details are given.

- **Chapter 4** deals with the most difficult aspect of EBM – how to get evidence into practice. Chapter 1 noted that EBM offers a challenge to traditional medical practice, and the professional debate about its validity has often been heated. Critics do not simply base their opposition on a desire to preserve tradition, but express fundamental doubts about the practicality and wisdom of what they see as 'diagnosis by scientific literature'. Though the EBM community defends itself vigorously against charges of 'cook book' medicine, not all are convinced. Moreover, those who support EBM may find the practical difficulties of following its principles all but insuperable. Chapter 4 looks at some of the strategies being implemented in pursuit of a clinically effective health service, including developments in evidence-based medical education and training, and in clinical audit and clinical guidelines. It ends by looking at the role of libraries and the information professional in supporting EBM which places information searching, evaluation and synthesis at the heart of good medical practice.

- **Chapter 5** looks at the patient perspective on EBM in the context of broader efforts to educate, inform and empower consumers of health care. Properly evaluated evidence presented in patient-friendly formats is seen by many as a key factor in redressing the traditional imbalance of power between doctor and patient, and (hopefully) encouraging empowered patients to make appropriate choices about treatment. However, doubts remain about the genuineness of empowerment and choice, and about how far patients can be educated or informed to make rational, evidence-based decisions. The patient's perspective on disease is inevitably different from that of the health care professional since only he or she experiences direct suffering, and what the doctor may see as a rational response to a particular condition may be very far from what the patient either wants or needs. Despite this potential barrier to the development of an evidence-based health service, the NHS is clearly committed to patient information, education and empowerment and sees patients as 'powerful agents for change' and sustained improvement in clinical effectiveness.

This review comes to no general conclusions about the likely success of EBM in transforming health care practice and the clinical/cost-effectiveness of the NHS. It is too early for that. However, it does illustrate the complexity of the issues and the multitude of cultural and organisational pitfalls that lie in wait for what seems – on the surface – such an obviously sensible notion. One thing does seem certain, that it will be many years before the principles of EBM become truly embedded within the health service. New generations of health care practitioners imbued by their professional education with a desire to learn, and equipped with the skills to search out and evaluate evidence, will need to become established and influential within the health service before EBM can really be said to have arrived. Equally, changes will need to be made to the environment within which they work to ensure that performance incentives promote, rather than stifle, the use of these skills. In the meantime, the champions of EBM need to guard its patient–centred virtues zealously if it is not to end up as just another weapon in the managerial cost-cutting armoury.

Acknowledgements

Reviewing the literature in any subject area is never an easy matter, and makes one only too aware of the deficiencies of the bibliographic databases and other sources that are used to identify potentially relevant material. In producing this guide I would like to acknowledge the invaluable help of Howard Jones of the British Library Science Reference and Information Service (SRIS) who has provided bibliographic support and much useful information from Internet sources. Thanks are also due to Trevor Sheldon of the NHS Centre for Reviews and Dissemination, and to Richard Wakeford and Bruce Madge of SRIS, for comment on drafts; to Sue Spedding and staff from the Library's Document Supply Centre (DSC) at Boston Spa for providing copies of many articles and reports; to Les Wilkinson of SRIS for additional bibliographic support; and to Mrs D Madge for the subject index.

Lesley Grayson
May 1997

1. WHAT IS EVIDENCE-BASED MEDICINE?

The current frenzy of interest in evidence-based medicine (EBM) within the health care world was initially prompted by the desire of a group of doctors to improve the quality of care provided to their patients, although it is unlikely that many of those patients are aware of what may turn out to be a revolution in the practice of medicine. Indeed, most would probably react to the notion of EBM with bewilderment. In the popular mind medicine is a 'scientific' activity, carried out by doctors and other health care professionals who have received a long and detailed scientific training and use highly sophisticated technologies to diagnose and treat illness. Surely those whose job it is to mend and maintain the human machine do so on the basis of rigorously evaluated scientific evidence? After all, we live in the modern world, not in the days of witch doctors and faith healers.

The reality is often rather different. The human being may be likened to a machine, but it is an extremely complex one about whose workings we still know relatively little. Moreover, its effective functioning is not simply a matter of making sure that the physical components work properly. The psyche and emotions of the individual, which are affected by an almost infinite number of often uncontrollable factors, also exercise a powerful influence over health and the effectiveness of medical treatment. Suffering and physical distress are not always the same thing, and the relief of suffering and the cure of disease are twin obligations of the medical profession. The failure to understand this vital distinction can result in medical intervention which, though technically correct and scientifically proven, may not only fail to relieve suffering but can even contribute to it[1].

A quote from one of the giants of EBM, Archie Cochrane, helps to illustrate the complexity of the situation which faces the medical practitioner. Although it relates to unusual circumstances – the treatment of a dying Soviet soldier in a wartime prison camp – it serves to underline the fact that science or factual knowledge is only one element of good medical practice, and may not always be the most important.

> 'The ward was full, so I put him in my room as he was moribund and screaming and I did not want to wake the ward. He had obvious gross bilateral cavitation and a severe pleural rub. I thought the latter was the cause of the pain and screaming. I had no morphia, just aspirin, which had no effect. I felt desperate. I knew very little Russian then and there was no-one in the ward who did. I finally instinctively sat down on the bed and took him in my arms, and the screaming stopped almost at once. He died peacefully in my arms a few hours later. It was not the pleurisy that caused the screaming, but loneliness. It was a wonderful education about the care of the dying. I was ashamed of my mis-diagnosis and kept the story secret'[2].

Cochrane's story suggests that scientific approaches to the practice of medicine are not necessarily the most appropriate. Moreover, the standards of scientific proof that one might expect in the more exact areas of human knowledge are not always possible in medicine. Where the units of study (patients) are so variable, it can be difficult to design rigorously controlled experiments to determine the effectiveness of a particular treatment. Even where this is possible, the results may be of limited practical value because few patients in the general population will match the exact characteristics of those chosen for the experiment[3].

Lay members of the public may be startled by widely quoted research from the US Office of Technology Assessment and other reputable sources which claims that only 10% to 20% of medical interventions are supported by solid scientific evidence[4]. Although this has been challenged more recently in two small scale studies which show much higher levels of use of scientifically validated treatments, there is still little doubt that many areas of medical practice continue to be based on habit, protocol or tradition rather than the most up-to-date research information[5]. David Weatherall in his book on *Science and the quiet art* notes that many medical specialties are still an uneasy mixture of solid science and unproved practice[6], while a comment from Rudolf Virchow nearly 150 years ago may still be valid today: 'it has been said that the rift between scientific and general medicine is so great that one can hold that the scholarly physician does nothing and the practical physician knows nothing'[7].

Resistance to change

However, only the most cynical observer would claim that medical practice is entirely unscientific – or, as one commentator has put it, the result of 'direct communication with God or...tossing a coin'. It has always involved a process of observation followed by assessment of possible treatment options – does it work, is it safe and (increasingly) is it worth it? The issue at stake is less whether doctors use evidence in their day-to-day work – clearly they do – but the breadth, depth and objectivity of the information on which they rely. Many doctors, like people in other walks of life, have a relatively narrow perspective, drawing largely on personal experience and that of a few colleagues when making decisions about patients. As a result, according to the EBM community, their levels of professional competence decline progressively: useless, wasteful and even dangerous practices continue against all the recommendations of reputable research, while new ones of proven efficacy are ignored[8,9].

A notable historical example of the latter relates to the treatment of scurvy. Although the effectiveness of lemon juice as a preventive treatment was clearly demonstrated as early as 1601, it was two hundred years before lemons were routinely included as part of ships' food stocks by the Navy. One might attribute the slowness of this particular knowledge transfer to the inadequate processes of information dissemination in the seventeenth and eighteenth centuries, but examples of resistance to new and scientifically proven treatments can still be found in the era of mass scientific publishing, online databases and the Internet. They include the slow adoption of corticosteroid therapy to improve lung function in cases of threatened premature birth, despite clear evidence that it reduces both neonatal morbidity and mortality.

For instances of the medical profession clinging to the past, one might point to the long continued use of dilatation and curettage (D&C) in response to dysfunctional uterine bleeding against evidence which shows that the procedure is frequently therapeutically useless and diagnostically inaccurate. D&C is one of the ten most frequently performed operations in the UK, of which only hernia repair and cataract removal are widely regarded as sound procedures with a high degree of clinical agreement about who will benefit and when they should be carried out. More questionable procedures include hysterectomy, tonsillectomy and the insertion of grommets to treat glue ear (reputable evidence now shows that some 80% of glue ear cases clear up without treatment).

Despite the, by now well publicised, reluctance of some members of the medical profession to move with the times, there are also examples of new treatments and procedures being adopted prematurely before adequate evaluation or training of those who carry them out. Keyhole surgery is one, in which inappropriate use or poor surgical skills have resulted in long term damage to a number of patients. Another is the use of chorionic villus sampling (CVS), heralded as a major advance in pre-natal diagnostic testing which could be performed much earlier in a pregnancy than existing tests such as amniocentesis. This was widely reported in the press and adopted for routine use in part, it is alleged, because of patient demand. Subsequent evaluation showed a link between the use of CVS before the tenth week of pregnancy and lower limb defects in babies, and it is now restricted to later use. Obstetrics has a particular reputation for adopting new techniques – for example, mass foetal screening – without good evidence of their benefit, and it seems that the medical profession is just as susceptible to high tech hype as the rest of us.

Information overload

These apparently irrational behaviours can still, in part, be attributed to inadequate knowledge transfer, compounded in the twentieth century by information overload caused by the massive increase in medical research and associated publication. Keeping track of new material, identifying what may be relevant, and evaluating it systematically are time consuming and skilled occupations for professionals who may already be fully stretched caring for patients and undertaking an increasing load of administrative duties. A tongue-in-cheek contribution to the *British Medical Journal* in Christmas week 1995 suggested that pressure to keep up with the literature can also produce physical damage ('polythenia gravis' – caused by slipping on piles of unread journals still in their plastic wrappers) and increasingly debilitating feelings of guilt and inadequacy[10]. Research from the USA claims that clinicians may need access to new, clinically important evidence as often as twice for every three patients seen, affecting up to eight clinical decisions a day[11]. In general medicine, it has been estimated that a well-briefed consultant would need the time to review 19 new articles every day of the year, a seemingly impossible task given the fact that most can devote less than an hour a week to keeping up with the literature[12].

The development of uniform requirements for manuscripts and for the reporting of randomised controlled trials (RCTs) may help[13]. So also will improvements in the abstracts of primary research articles[14]. However, in practice, most people will need to rely on the reviewing skills of others. Unfortunately, the quality of many reviews leaves much to be desired[15]. A significant proportion of primary research in the medical field is of dubious scientific quality, lacking 'either relevance or sufficient methodological rigour to be reliable enough for answering clinical questions'[16]. Some research, including double-blind RCTs, may be corrupt[17] while publication bias[18] and the under-reporting of research[19] are also significant problems. However, reviewers often fail to approach their task in a systematic way, missing effective new research while continuing to recommend other forms of treatment after reputable controlled trials have shown them to be ineffective or even, on occasion, actively harmful[20].

The creation of an international registry of clinical trials is widely seen as a necessary prerequisite of exerting greater control and permitting the systematic review of both published and unpublished RCTs, and this is now in progress under the aegis of the Cochrane Collaboration[21]. The difficulties faced by reviewers in identifying relevant studies have been ameliorated to some extent by changes in indexing practice for the MEDLINE and EMBASE (Excerpta Medica) databases, but major problems still remain[22]. Not the least of these is the fact that MEDLINE indexes less than 4,000 of the some 15,000 biomedical journals received by the National Library of Medicine. When one adds to this the fact that a significant proportion of RCTs may never reach the published stage at all, it is clear that it will be some time before the registration task is accomplished. In the interim, many practitioners will remain in ignorance of good quality research evidence and/or doubtful of its practical value.

At the same time, there is an inherent conservatism in medicine as in most areas of human activity, especially those which require high levels of skill and long periods of training. Achievement in medicine demands a level of dedication to hard work and study over many years which few of us experience, and it is hardly surprising that those who succeed often have a high degree of confidence in their knowledge and ability to solve any clinical problem with which they are faced. Equally, they may sometimes be resistant to changes which cast doubt on that knowledge and ability, and the self-esteem which goes with them. The professional debate on the appropriateness of EBM can, as Chapter 4 shows, be heated and those who seek to spread its principles can often face considerable hurdles.

However, the increasing pace of medical technological advance means that sole reliance on a combination of early training, experience and the advice of colleagues is no longer felt to be an appropriate basis for medical practice. The pressure for EBM which has grown within sectors of the medical profession in recent years is a direct response to the belief that basing clinical decisions on opinion, precedent, the recommendations of personal colleagues or (worst of all) those of drug company representatives is no longer acceptable, particularly in the context of increasing levels of awareness and knowledge among the patient population. However difficult it may be to provide scientific standards of proof in some areas of medicine, an approach which links research evidence to clinical practice has to be better than reliance on personal experience and prejudice. Not only should it eliminate ineffective treatments and promote more effective ones, to the benefit of both the patient and the taxpayer, it should also help to ensure greater continuity and uniformity of care across the health service through the development of common approaches and best practice guidelines.

The advocates of EBM

In the UK the single most influential figure in the development of evidence-based medicine was Archie Cochrane, former director of the Medical Research Council's epidemiology unit and the first president of the British Faculty of Community Medicine. His earliest thoughts on EBM, as already noted, were prompted by his experiences in prisoner-of-war camps where he conducted his first controlled trial on the effectiveness of yeast supplements in treating nutritional oedema. Perhaps even more importantly for the future of the EBM movement, he also noted that despite the fact that he was unable to provide seriously ill patients with the treatments he had learned about during training, many recovered anyway. He returned to normal medical practice with a strong desire for more reliable information about which treatments helped patients, and which did not.

Cochrane's book on *Effectiveness and efficiency*, originally published in 1972, stated in simple, straightforward terms that resources for health care will always be limited and that they should be used to provide equitably those forms of health care which have been shown to be effective through properly designed evaluations[23]. Cochrane's gold standard of evaluation was the randomised controlled trial, a rarity in medical research before the 1960s but now universally applied to the investigation of new drug therapies and increasingly to surgical therapies and diagnostic tests. The systematic identification and appraisal of RCTs in specific fields has subsequently become a key element in EBM, designed to synthesise the findings of a range of reputable research studies using the statistical techniques of meta-analysis[24,25].

Before he died in 1988, Archie Cochrane saw the beginnings of a systematic review of RCTs in some areas of health care including gynaecology and obstetrics. This particular sector of the medical community may have been finally stung into action by Cochrane's own relentless campaign which included the award of a wooden spoon to mark its reluctance to embrace EBM principles. He did, however, contribute the foreword to *Effective care in pregnancy and childbirth* – a compendium of systematic reviews which 'has shaken obstetrics world-wide' – when it eventually appeared in 1989[26].

In 1992 a centre bearing his name was founded as part of the new National Health Service R&D programme, and led to the astonishingly rapid development of the Cochrane Collaboration as a world-wide initiative preparing, maintaining and disseminating systematic, up-to-date reviews or meta-analyses of RCTs and, where these are not available, the most reliable evidence from other sources[27]. The Cochrane Collaboration, as noted in Chapter 3, is a voluntary initiative in which participants commit themselves to the review of research evidence in their fields of interest for the remainder of their working life. As such, it reflects Archie Cochrane's passionate commitment to the public service ideals of the NHS and to the interests of the patient. The Collaboration's logo is an abstract representation of the systematic review of RCTs in the field of corticosteroid therapy for threatened premature birth – a review which, had it been carried out earlier, could have prevented the serious damage or death of tens of thousands of babies.

In more recent years Cochrane's role as the guru of EBM has been taken over by the American David Sackett who has spent most of his working life in researching the potential of applying epidemiological and biostatistical ways of thinking to the care of individual patients. In the early 1960s he became a field epidemiologist in the US Public Health Service and, as his skills in this area developed, 'it dawned on him that applying these epidemiologic principles (plus a few more from biostatistics) to the beliefs, judgments, and intuitions that comprise the art of medicine might substantially improve the accuracy and efficiency of diagnosis and prognosis, the effectiveness of management, the efficiency of trying to keep up to date, and, of special importance, the ability to teach others how to do these things'[28].

Later in the 1960s, he helped to found the Department of Clinical Epidemiology and Biostatistics at McMaster University in Canada and pioneered the development of problem-based, self-directed education both for undergraduate medical students and for older doctors steeped in the attitudes and approaches of the traditional model of medical practice[29]. In 1994 he moved to Oxford to a new Chair in Clinical Epidemiology at the Nuffield Department of Medicine, and founded the UK's first Centre for Evidence-Based Medicine which is rapidly becoming a key focus for promoting the teaching, learning, practice and evaluation of EBM throughout the UK[30].

A challenge to authority

Although the views of Archie Cochrane, David Sackett and others have struck a chord with many lay people, health service professionals and politicians, their translation into practice has been slow. EBM offers a radical challenge to the time-honoured model of medical practice which places a high value on traditional scientific authority and adherence to standard approaches. In David Sackett's words, this model assumes that:

- 'Unsystematic observations from clinical experiences are both valid and sufficient for building and maintaining one's knowledge about patient prognosis, the value of diagnostic tests, and the efficacy of treatment.
- The study and understanding of basic mechanisms of disease and pathophysiologic principles is a sufficient guide for clinical practice.
- A combination of thorough traditional medical training and common sense is sufficient to allow one to evaluate reports of new tests and treatments.
- Content expertise and clinical experience are a sufficient base from which to generate valid guidelines for clinical practice.'[31]

Thus a doctor faced with a problem can solve it using a combination of clinical experience and knowledge of the underlying biology backed up, if necessary, by reference to the relevant textbook. If still stumped, he or she can refer to a respected colleague or check the recognised authorities in the medical literature.

The evidence-based approach to medicine places a much lower value on scientific authority and adherence to standard approaches although the importance of traditional knowledge and skills, including sensitivity to the individual patient's emotional needs, is not denied. Again in Sackett's words, the assumptions of EBM are that:

- 'Clinical experience, and the development of clinical instincts (particularly with respect to diagnosis), are crucial and necessary parts of becoming a competent physician. Moreover, many aspects of clinical practice cannot, or will not, ever be adequately tested, and clinical experience, and its lessons, are particularly important in these situations. At the same time, systematic attempts to record observations in a reproducible and unbiased fashion markedly increases the confidence one can have in knowledge about patient prognosis, the value of diagnostic tests, and the efficacy of treatment. In the absence of systematic observation one must be cautious in the interpretation of information derived from clinical experience and intuition, for they have at times been highly misleading.
- The study and understanding of basic mechanisms of disease are necessary but insufficient guides for clinical practice. The rationales for diagnosis and treatment which follow from basic pathophysiologic principles are often incorrect, leading to inaccurate predictions about the performance of diagnostic tests and the efficacy of treatments.
- Understanding certain rules of evidence is necessary to correctly interpret literature on causation, prognosis, diagnostic tests, and treatment strategy.'

In brief, evidence-based medical practice can be described as 'the conscientious, explicit and judicious use of current best evidence in making decisions about the care of individual patients', based on skills which allow the doctor to evaluate both personal experience and external evidence in a systematic and objective manner[32].

The EBM community often describes this as a new 'paradigm', or a revolution in traditional medical practice, an assertion which irritates those many members of the profession who insist that doctors have always evaluated evidence in their day-to-day work. In response, the EBM community has begun to moderate its language and now tends to refer to EBM as an enhancement of traditional skills rather than a replacement for them. David Sackett, in his January 1996 *British Medical Journal* editorial, argues that 'the practice of evidence-based medicine means integrating individual clinical expertise with the best available external clinical evidence from systematic research...neither alone is enough'.

The decreased emphasis on authority within EBM should not, in Sackett's view, be taken as a rejection of what can be learned from the experience of teachers and colleagues. Nor does EBM discount the traditional skills of medical training. A sound knowledge of the underlying pathophysiology, and sensitivity to the patient's emotional needs, remain crucial. However, additional skills are also required to equip doctors to define problems and to track down, evaluate and apply evidence from the medical research base. Though these skills enhance (rather than supplant) tradition, Sackett continues to insist that the demonstrable failings of current medical practice mean that the arguments of those who claim that doctors have always practised EBM are untenable.

Overcoming scepticism

Although EBM seems to many an intuitively commonsense approach to medical practice, some doctors remain doubtful about its desirability and feasibility. This is not simply the result of conservative defensiveness although, as Chapter 4 suggests, this certainly plays a part. There are also serious intellectual and cultural misgivings about the philosophy of EBM among those doctors who view the practice of medicine as an art just as much as a science. Some of these may arise from misunderstandings about what EBM means, hence (no doubt) the increasing emphasis of David Sackett and others on its role in building on the strengths of traditional medical practice. However, worries remain about the wisdom of placing too much reliance on RCTs or other epidemiological evidence in the treatment of individual patients. In addition, there are practical difficulties involved in dealing with information overload and ensuring that relevant information is delivered at the time it is needed and in a format that is useful in the treatment of the patient.

The scale of the difficulties to be overcome in its successful implementation means that it may be all too easy for the sceptic to dismiss EBM as just another fad and retreat into the comfort of the traditional model of practice. Many doctors face a burdensome workload and find it difficult, as do the rest of us, to learn new ways of doing things. Even the simplest new treatment may require them to remember a whole host of facts about indications, contra-indications, drug interactions, dosages etc., perhaps making the conventional options doubly attractive. It is clear that the single most important issue to be resolved in EBM implementation is not the creation, collection and evaluation of good quality evidence or even its physical dissemination to those who might make use of it. Difficult though these issues are, they are largely technical. Much more problematic, as Chapter 4 discusses, will be solving the perennial difficulty of ensuring that new information is translated into everyday practice.

This can, of course, be approached through education – both of medical students and practising doctors – and through the development of effective and imaginative ways of presenting and delivering information via a range of mechanisms including the Internet. The role of health care information professionals in acting as information filters or gatekeepers to the many information sources available is also crucial. These routes to EBM can help to develop skills and change behaviour from within. However, in practice, relatively few currently practising doctors and other health care professionals are likely to experience EBM as a purely self-directed enhancement of their professional skills. Far more are likely to have it thrust upon them through the medium of clinical audit, clinical guidelines and other management-driven initiatives which may be welcomed if they are developed in partnership with clinical staff, but could equally be seen by some as a potential constraint on their traditional freedom of action.

As the next chapter discusses, the principles of EBM – in particular its potential for making choices between more and less effective treatments – have been enthusiastically embraced by both politicians and health service managers as a way of addressing the problems of a resource-constrained health service. EBM is no longer solely a means for the individual doctor to improve the quality of care provided to his or her patients, but has taken on another role as a tool for exercising control and influence within the health service through clinical guidelines and other mechanisms. David Grahame-Smith, in his 'Socratic dialogue' published in the *British Medical Journal* in April 1995, puts it thus: 'The main barrier they [the health service managers] perceive is an anarchic medical profession spending their money in a profligate and unnecessary manner. They see your beloved evidence-based medicine as a means to shackle the doctors and bend them to their will. That, I am certain, is why they are so enthusiastic about it'[33]. Not, perhaps, a scenario which Archie Cochrane envisaged in his long campaign for an effective and socially just health service, but one which may well have significant implications for the way this new approach to health care is received by those who are to implement it.

(1)
THE NATURE of suffering and the goals of medicine
Cassell, E J
New England Journal of Medicine, 18 Mar 1982 306(11) pp639-45
Argues that the issue of suffering and its relationship to organic illness is rarely discussed in the medical literature, and offers a description of the nature and causes of the former. Develops a distinction between suffering and physical distress based on clinical observations, noting that the former is experienced by persons not just bodies. Thus it can include physical pain but is not limited to it, and the relief of suffering and the cure of disease are twin obligations of a truly caring medical profession.

(2)
ONE man's medicine: an autobiography of Archie Cochrane
Cochrane, A L; Blythe, M
BMJ (Memoir Club): London, 1989. 283pp
For an account of Cochrane's wartime controlled trial involving yeast supplements
See also: Sickness in Salonica: my first, worst and most successful clinical trial, by A L Cochrane (British Medical Journal, 22-29 Dec 1984 289(6460) pp1726-27)

(3)
HIGH technology medicine: benefits and burdens
Jennett, B
Nuffield Provincial Hospitals Trust, 59 New Cavendish Street, London W1M 7RD, 1984. 245pp
Doubts the universal appropriateness of RCTs as a means of evaluating a process which is often more of an art than a science. Also questions the practical value of information

derived from trials in which patients are very carefully selected, for example to exclude other conditions. Such trials may produce results which are scientifically rigorous but of limited practical value because they relate to atypical conditions.

(4)
WHERE is the wisdom...? The poverty of medical evidence
Smith, R
British Medical Journal, 5 Oct 1991 303(6806) pp798-99
Summarises the debate which alleges that only a relatively small proportion of medical treatments are scientifically validated. For the widely quoted assessments by the Office of Technology Assessment, and for a later analysis by the US National Institutes of Health
See also: Assessing the efficacy and safety of medical technologies, by Office of Technology Assessment (US Government Printing Office, 1978. 133pp)
The impact of randomised clinical trials on health policy and medical practice, by Office of Technology Assessment (US Government Printing Office, 1983. 109pp)
Analysis of the National Institutes of Health Medicare coverage assessment, by M Dubinksy and J H Ferguson (International Journal of Technology Assessment in Health Care, 1990 6(3) pp480-88)
Note: The Office of Technology Assessment has been abolished by the US Congress, but the text of the two reports listed above, and others relating to health care issues, can be accessed via **http://www.wws.princeton.edu:80/~ota**. A complete collection of OTA reports has also been issued on a set of five CD-ROMs and is available from the US Government Publishing Office, PO Box 371954, Pittsburgh, PA 15250-7974.

(5)
INPATIENT general medicine is evidence based
Ellis, J and others
Lancet, 12 Aug 1995 346(8972) pp407-10, 37 references
Questions the widely quoted figures suggesting that only 10%-20% of medical interventions are based on scientifically reputable evidence. Presents the findings of a study of a general medical team at a university-affiliated general hospital in Oxford which looked at the treatments given to 109 patients, and checked to see if there was RCT evidence that they were effective. 82% were classified as evidence-based, i.e. there was RCT support (53%) or unanimity in the clinical team about the existence of convincing non-experimental evidence (29%). Although the study needs to be repeated in other clinical settings, it suggests that pessimism over the degree to which EBM is already practised is misplaced. For letters in response to this article, and for a later study from the Leeds University Centre for Research in Primary Care which claims very similar levels of use of sound evidence in general practice.
See also: Evidence-based medicine [letters] (Lancet, 23 Sep 1995 346(8978) pp837-40)
Evidence-based general practice: a retrospective study of interventions in one training practice, by P Gill and others (British Medical Journal, 30 Mar 1996 312(7034) pp819-21)

(6)
SCIENCE and the quiet art: the role of medical research in health care
Weatherall, D J
Oxford University Press, 1995. 378pp
An analysis commissioned by the Commonwealth Fund, and directed at the lay reader. Looks at the successes and failures of modern medicine, noting that practice remains rooted in a mix of science and experience. For more comment from David Weatherall

See also: The inhumanity of medicine: time to stop and think, by D J Weatherall (British Medical Journal, 24–31 Dec 1994 309(6970) pp1671–72)

(7)
RUDOLF Virchow (translated by S G M Engelhardt)
In: Concepts of health and disease: interdisciplinary perspectives, edited by A L Caplan and others (Addison-Wesley: Reading, MA, 1981 pp187–95)
Rudolf Virchow (1821–1902) was Professor of Pathological Anatomy at Berlin University in the mid-nineteenth century. Three translated selections from his published work are included in Section 2.5 of Part 2 of this compilation including 'Concerning standpoints in scientific medicine' originally published in *Archiv für Pathologische Anatomie und Physiologie und für Klinische Medicin* (1847, 1 pp3–7) which Virchow founded. In it, Virchow argues that scientific medicine has as its object the 'discovery of changed conditions', while practical medicine has as its object the healing of patients. As a result 'medicine in practice is thus never scientific medicine itself, even when it is in the hands of the greatest master, but rather only an application of it'. Virchow regrets the emergence of a 'rift between science and practice' though he claims that this has not happened in English medicine.

(8)
A COMPARISON of results of meta-analyses of randomized control trials and recommendations of clinical experts: treatments for myocardial infarction
Antman, E M and others
Journal of the American Medical Association, 8 Jul 1992 268(2) pp240–48
Compares the evidence accumulating for RCTs and systematic reviews of treatments for myocardial infarction with the recommendations made in contemporaneous textbooks. Finds that most texts continue to recommend treatments which scientific evidence has shown to be useless, while failing to recommend new proven therapies. Also finds that most reviews of evidence in this area fail to take into account the results of strong evidence that has emerged in systematic reviews of relevant RCTs.

(9)
CHANGES over time in the knowledge base of practising internists
Ramsey, P G and others
Journal of the American Medical Association, 28 Aug 1991 266(8) pp1103–07
Presents the results of a study to determine factors affecting the knowledge base of 289 practising internists who had received American Board of Internal Medicine certification in the previous 5 to 15 years. Participants were tested using a multiple choice examination with questions from the 1988 ABIM examination. The results show a significant inverse correlation between examination scores and the number of years elapsed since certification, with knowledge declining sharply within 15 years of certification.

(10)
POLYTHENIA gravis: the downside of evidence based medicine
Down End Research Group
British Medical Journal, 23–30 Dec 1995 311(7021) pp1666–68
Reviews, on the basis of impeccable scientific evidence, the physical and mental trauma suffered by doctors in their vain attempts to keep up with the literature in their specialties. Polythenia gravis (caused by slipping on piles of unread journals still in their wrappers) and the guilt and feelings of inadequacy which often accompany it, are rapidly relieved by donating all journals, still unread, to needy doctors in Third World countries.

(11)
INFORMATION needs in office practice: are they being met?
Covell, D G and others
Annals of Internal Medicine, Oct 1985 103(4) pp596-99
Studies the self-reported information needs of 47 physicians during a half day of typical surgery practice. During this period physicians raised 269 questions relating to patient management, with only 30% of information needs met during the patient visit, usually by a colleague. Print sources were rarely used because textbooks were out of date, journal articles were poorly organised, and books and drug information sources poorly indexed. In addition there was a lack of knowledge of appropriate sources of information, and a lack of time to use them.

(12)
EVIDENCE-based medicine: a new journal to help doctors identify the information they need
Davidoff, F and others
British Medical Journal, 29 Apr 1995 310(6987) pp1085-86, 8 references
An editorial introducing the journal *Evidence-Based Medicine* which discusses the information problems facing doctors. It notes the fact that a practitioner in adult internal medicine would, theoretically, have had to read over 6,000 articles during 1992 in order to be fully up to date with new, clinically important research. For more from Frank Davidoff
See also: Evidence-based medicine: why all the fuss? by F Davidoff and others (Annals of Internal Medicine, 1 May 1995 122(9) p727)

(13)
UNIFORM requirements for manuscripts submitted to biomedical journals
International Committee of Medical Journal Editors
Journal of the American Medical Association, 5 May 1993 269(17) pp2282-86
The International Committee, known as the Vancouver Group, was set up in 1978. It produces guidance on many aspects of publishing in the biomedical field which is widely followed by journal editors. For the recent CONSORT statement which lays down agreed rules on the reporting of RCTs
See also: Improving the quality of reporting of randomized controlled trials: the CONSORT statement, by C B Begg and others (Journal of the American Medical Association, 28 Aug 1996 276(8) pp637-39, 31 references + editorial endorsement from JAMA editor Drummond Rennie on p649)
Better reporting of randomised controlled trials: the CONSORT statement, by D G Altman (British Medical Journal, 7 Sep 1996 313(7057) pp570-71, 12 references)
The CONSORT statement [letters] (Journal of the American Medical Association, 18 Dec 1996 276(23) pp1876-77)

(14)
A PROPOSAL for more informative abstracts of clinical articles
Ad Hoc Working Group for Critical Appraisal of the Medical Literature
Annals of Internal Medicine, Apr 1987 106(4) pp598-604
Following proposals in 1987 and 1988, several medical journals have developed more structured and informative abstracts to facilitate pre-publication peer review, help clinical readers to identify relevant articles, and allow more precise computer-based literature searches. For more on this issue
See also: Structured abstracts for papers reporting clinical trials, by E J Huth (Annals of

Internal Medicine, Apr 1987 106(4) pp626-27)
A proposal for more informative abstracts of review articles, by C D Mulrow and others
(Annals of Internal Medicine, Apr 1988 108(4) pp613-15)
More informative abstracts revisited, by R B Haynes and others (Annals of Internal
Medicine, 1 Jul 1990 113(1) pp69-76)
More informative abstracts of articles describing clinical practice guidelines, by R S A
Hayward and others (Annals of Internal Medicine, 1 May 1993 118(9) pp731-37)
Quality of nonstructured and structured abstracts of original research articles in the *British
Medical Journal*, the *Canadian Medical Association Journal* and the *Journal of the American
Medical Association*, by A Taddio and others (Canadian Medical Association Journal, 15
May 1994 150(10) pp1611-15)

(15)
THE MEDICAL review article: state of the science
Mulrow, C D
Annals of Internal Medicine, Mar 1987 106(3) pp485-88
Assesses 50 reviews published between June 1985 and June 1986 on the basis of eight
criteria. Of the 50 reviews, 17 satisfied three of the criteria, 32 satisfied four or five, and
one satisfied six. Qualitative synthesis was often used to integrate information, with
quantitative synthesis rarely used. Concludes that current reviews do not routinely use
scientific methods to identify, assess and synthesise information, and that the methods used
in this assessment could beneficially be applied to reviews themselves. For guidelines issued
at around the same time to help practitioners distinguish between the good and the bad in
reviews, and for later comment on an index of quality
See also: Guidelines for reading literature reviews, by A D Oxman and G H Guyatt
(Canadian Medical Association Journal, 15 Apr 1988 138(8) pp697-703)
Validation of an index of the quality of review articles, by A D Oxman and G H Guyatt
(Journal of Clinical Epidemiology, 1991 44(11) pp1271-78)

(16)
EVIDENCE based medicine: an approach to clinical problem-solving
Rosenberg, W M; Donald, A
British Medical Journal, 29 Apr 1995 310(6987) pp1122-26, 10 references
Discusses EBM as a process of turning clinical problems into questions, and systematically
locating, assessing and using research evidence as a basis for decision making. EBM skills
can be taught to, and practised by, clinicians at all levels of seniority and will help to bridge
the gap between good clinical research and practice. EBM integrates medical education
and clinical practice, and has the potential for promoting continuity and uniformity of care
through the development of common approaches and guidelines. Concludes by listing
both the advantages and disadvantages of EBM.

(17)
FRAUD, deception and misconduct in science
Grayson, L
Science, Technology and Innovation, Feb 1996 9(1) pp16-22, 10 references
Reviews the factors which may induce researchers to cheat, beginning with comment on a
study by Kenneth Schulz which alleges very high levels of misconduct and poor
methodological standards in allegedly randomised drug trials. Also looks at the response of
the scientific community to misconduct, and at the policy implications. Based on a British
Library analysis of the literature of scientific misconduct, much of which concerns the

biomedical disciplines. For details of the Schulz study

See also: Empirical evidence of bias: dimensions of methodological quality associated with estimates of treatment effects in controlled trials, by K F Schulz and others (Journal of the American Medical Association, 1 Feb 1995 273(5) pp408-12)

Randomised trials, human nature, and reporting guidelines, by K F Schulz (Lancet, 31 Aug 1996 348(9027) pp596-98)

Subverting randomization in controlled trials, by K F Schulz (Journal of the American Medical Association, 8 Nov 1995 274(18) pp1456-58)

(18)
PUBLICATION bias in clinical research
Easterbrook, P J and others
Lancet, 13 Apr 1991 337(8746) pp867-72
Presents the results of a retrospective survey of 487 research projects approved by the Central Oxford Research Ethics Committee between 1984 and 1987. As of May 1990, 285 of the studies had been analysed and 52% published. Finds that studies with statistically significant results were more likely to be published than those which found no differences between study groups. Studies with significant results were also more likely to lead to a greater number of publications, and had an increased likelihood of publication in journals with a high citation impact. For more on this issue

See also: Confronting publication bias: a cohort design for meta-analysis, by R J Simes (Statistics in Medicine, Jan-Feb 1987 6(1) pp11-29)

Reference bias in reports of drug trials, by P C Gøtzsche (British Medical Journal, 12 Sep 1987 295(6599) pp654-56)

Publication bias and clinical trials, by K Dickersin and others (Controlled Clinical Trials, Dec 1987 8(4) pp343-53)

Publication bias: a problem in interpreting medical data, by C B Begg and J A Berlin (Journal of the Royal Statistical Society (A), 1988 151(3) pp419-63)

An assessment of publication bias using a sample of published clinical trials, by J A Berlin and others (Journal of the American Statistical Association, Jun 1989 84(406) pp381-92)

Methodology and overt and hidden bias in reports of 196 double-blind trials of nonsteroidal anti-inflammatory drugs in rheumatoid arthritis, by P C Gøtzsche (Controlled Clinical Trials, Mar 1989 10(1) pp31-56)

Publication bias and dissemination of clinical research, by C B Begg and J A Berlin (Journal of National Cancer Institute, 18 Jan 1989 81(2) pp107-15)

Minimising the three stages of publication bias, by T C Chalmers and others (Journal of the American Medical Association, 9 Mar 1990 263(10) pp1392-95)

Publication bias: the triumph of hope over experience, by D Rennie and A Flanagin (Journal of the American Medical Asociation, 15 Jan 1992 267(3) pp411-12)

NIH clinical trials and publication bias, by K Dickersin and others (Online Journal of Current Clinical Trials [serial online], 28 Apr 1993, Doc. No. 50)

(19)
UNDER-REPORTING research is scientific misconduct
Chalmers, I
Journal of the American Medical Association, 9 Mar 1990 263(10) pp1405-08, 38 references
Discusses the practice of failing to publish non-confirmatory data altogether, or to publish only those portions of evidence that substantiate the clinical or experimental hypothesis. For more on the under-reporting issue

See also: Factors influencing publication of research results: follow-up of applications submitted to two institutional review boards, by K Dickersin and others (Journal of the American Medical Association, 15 Jan 1992 267(3) pp374-78)

Should unpublished data be included in meta-analyses? Current convictions and controversies, by D J Cook and others (Journal of the American Medical Association, 2 Jun 1993 269(21) pp2749-53)

Completeness of reporting of trials published in languages other than English: implications for conduct and reporting of systematic reviews, by D Moher and others (Lancet, 10 Feb 1996 347(8998) pp363-66)

(20)

THE FUTURE of medical journals: in commemoration of 150 years of the British Medical Journal

Lock, S (editor)

BMJ Publishing: London, 1991. 217pp

Stephen Lock's introduction summarises the two major themes of a Ciba Foundation symposium on the future of biomedical journals: the scientific quality of journals, and the relevance of their content to clinicians. Finds, in general, that the health of the scientific journal is 'more apparent than real' and that journals serve the medical community poorly. Section III on the communication of clinical information is particularly relevant to the EBM debate, and includes contributions on 'The changing editorial paradigm: hard data are necessary, but no longer sufficient, conditions for publishing clinical trials', by P Riis (pp101-107); 'Closing the gap between what researchers can do and what clinicians use: the journals' role', by R H Fletcher (pp108-115); 'How clinical journals could serve clinician readers better', by R B Haynes (pp116-26); and 'Improving the quality and dissemination of reviews of clinical research', by I Chalmers (pp127-46).

(21)

MAKING clinical trialists register

Lancet, 27 Jul 1991 338(8761) pp244-45

For more on this issue, including details of the development of an international register by the Cochrane Collaboration

See also: Publication bias: the case for an international registry of clinical trials, by R J Simes (Journal of Clinical Oncology, Oct 1986 4(10) pp1529-41)

Towards prospective registration of clinical trials, by C L Meinert (Controlled Clinical Trials, Mar 1988 9(1) pp1-5)

Retrospective and prospective identification of unpublished controlled trials: lessons from a survey of obstetricians and pediatricians, by J Hetherington and others (Pediatrics, Aug 1989 84(2) pp374-80)

Creating an NIH clinical trials registry: a user-friendly approach to health care, by W R Harlan (Journal of the American Medical Association, 8 Jun 1994 271(22) p1729)

Establishing and maintaining an international register of RCTs, edited by K Dickersin and K Larsen (In: Cochrane Collaboration handbook, edited by D Sackett and A Oxman. Cochrane Collaboration: Oxford, 1995. Section V)

(22)

IDENTIFYING relevant studies for systematic reviews

Dickersin, K and others

British Medical Journal, 12 Nov 1994 309(6964) pp1286-91

Reviews research on this issue over the previous decade, noting that only some 77% of RCT reports indexed in MEDLINE can be reliably retrieved because of the lack of suitable descriptor terms and/or their inconsistent application by indexers. For more comment on the problems of identifying RCTs

See also: An investigation of the adequacy of MEDLINE searches for randomized controlled trials (RCTs) of the effects of mental health care, by C E Adams and others (Psychological Medicine, Aug 1994 24(3) pp741-48)

The Cochrane Collaboration: the role of the UK Cochrane Centre in identifying the evidence, by C Lefebvre (Health Libraries Review, Dec 1994 11(4) pp235-42, 18 references)

Identifying reports of controlled trials in the *BMJ* and the *Lancet*, by S J McDonald and others (British Medical Journal, 2 Nov 1996 313(7065) pp1116-17)

(23)
EFFECTIVENESS and efficiency: random reflections on health services
Cochrane, A L

Nuffield Provincial Hospitals Trust, 59 New Cavendish Street, London W1M 7RD, 1972. 103pp (Reprinted in association with the British Medical Journal in 1989)

For Cochrane's views five years later, in which he lamented that 'it is surely a great criticism of our profession that we have not organised a critical summary, by specialty or subspecialty, adapted periodically, of all relevant randomized controlled trials'

See also: 1931-1971: a critical review, with particular reference to the medical profession, by A L Cochrane (In: Medicines for the year 2000. Office of Health Economics, 12 Whitehall, London SW1A 2DY, 1979 pp1-11)

(24)
META-ANALYSIS in clinical research
L'Abbé, K A and others

Annals of Internal Medicine, Aug 1987 107(2) pp224-33, 73 references

An often quoted article on the techniques of meta-analysis, with comment on issues such as protocol development, objectives, literature searching, publication bias, measures of study outcomes and quality of data. Includes guidelines to permit the assessment of meta-analysis quality.

(25)
SYSTEMATIC reviews
Chalmers, I; Altman, D G (editors)

BMJ Publishing: London, 1995. 119pp

Brings together revised versions of articles published in the *British Medical Journal* and also includes a classified bibliography of references about systematic reviews, pp96-112. A useful introduction to the issues and problems involved in the systematic review of RCTs, a complex process which has generated a large literature of its own. No attempt is made in this review to cover this material in detail. However, for a selection of other reviews of systematic review and meta-analysis

See also: Summing up: the science of reviewing research, by R J Light and D B Pillemer (Harvard University Press: Cambridge, MA, 1984. 191pp)

Meta-analysis: quantitative methods for research synthesis, by F M Wolf (Sage Publications: Beverly Hills, CA, 1986. 65pp. Sage University Paper on Quantitative Applications in the Social Sciences, Series No. 07-059)

The handbook of research synthesis, edited by H Cooper and L V Hedges (Russell Sage Foundation: New York, 1994. 576pp)

Meta-analysis: decision analysis and cost-effectiveness analysis: methods for quantitative synthesis in medicine, by D B Petitti (Oxford University Press: New York, 1994. 256pp. Monographs in Epidemiology and Biostatistics, Vol. 24)

Undertaking systematic reviews of research on effectiveness: CRD guidelines for those carrying out or commissioning reviews (NHS Centre for Reviews and Dissemination, University of York, Hesington, York YO1 5DD, Jan 1996. 104pp)

Conference on meta-analysis in the design and monitoring of clinical trials, edited by Y J Lee (Statistics in Medicine, Jun 1996 15(12). Complete issue reporting a 1994 conference)

(26)
EFFECTIVE care in pregnancy and childbirth
Chalmers, I and others (editors)
Oxford University Press, 1989. 2 volumes (1,516pp)
A compendium of meta-analyses which has 'shaken obstetrics world wide' not only for what it shows to be true, but for the many clinical strategies on which it casts doubt. Including a foreword by Archie Cochrane, this book was instrumental in the development of the international Cochrane Collaboration. Now updated electronically as well as in hard copy, with new editions in 1991 and 1995. For the companion volume on neo-natal care which followed, and for recent comment on the need for an evidence-based approach to obstetrics
See also: Effective care of the newborn infant, edited by J C Sinclair and M B Bracken (Oxford University Press, 1992. 650pp)
The need for evidence-based obstetrics, by M W Enkin (Evidence-Based Medicine, Jul/Aug 1996 1(5) pp132-33)

(27)
GETTING to grips with Archie Cochrane's agenda: all randomized controlled trials should be registered and reported
Chalmers, I and others
British Medical Journal, 3 Oct 1992 305(6857) pp786-88, 25 references
The Director of the UK Cochrane Centre discusses the need for systematic review of RCTs, arguing that the fundamental challenge is to identify all potentially relevant studies. Existing methods, including the use of bibliographic databases, identify only a proportion, and there is a need for the registration of all RCTs at inception. For more on this theme
See also: Cochrane's legacy (Lancet, 7 Nov 1992 340(8828) pp1131-32)

(28)
CLINICAL epidemiology: a basic science for clinical medicine: 2nd edition
Sackett, D L and others
Little, Brown: Boston, 1991. 441pp
David Sackett's book, written with Brian Haynes, Gordon Guyatt and Peter Tugwell, begins with a preface outlining the background to their interest in applying epidemiological ways of thinking to clinical practice, and summarising the reasons why a second edition became necessary. It includes three sections on diagnosis, management and keeping up to date, the last including discussion of how to review personal performance, how to track down evidence to solve clinical problems, how to survey the medical literature, how to read reviews and economic analyses, and how to get the most from (and give the most to) continuing medical education.

(29)

EVIDENCE-based medicine: a new approach to teaching the practice of medicine

Guyatt, G H and others

Journal of the American Medical Association, 4 Nov 1992 268(17) pp2420-25, 37 references

The Evidence-Based Medicine Working Group, including David Sackett, discusses the 'paradigm shift' from the traditional practice of medicine to the new evidence-based approach; the skills required for the practice of EBM; and possible barriers to its adoption. Includes comment on the teaching of EBM at McMaster University in Canada. For letters in response to this article

See also: Evidence-based medicine: a new paradigm [letters] (Journal of the American Medical Association, 10 Mar 1993 269(10) pp1253-54)

(30)

MCMASTER'S pioneer in evidence-based medicine now spreading his message in England

Cohen, L

Canadian Medical Assocation Journal, 1 Feb 1996 154(3) pp388-90

Looks at David Sackett's achievements as a proponent of EBM through his own views and those of colleagues. For his thoughts on some of the dilemmas facing the medical practitioner

See also: The doctor's (ethical and economic) dilemma: a description of the dilemmas faced by a physician who tries to serve both individual patients and society, by D L Sackett (Office of Health Economics, 12 Whitehall, London SW1A 2DY, Aug 1996. 22pp)

(31)

APPLYING overviews and meta-analyses at the bedside

Sackett, D L

Journal of Clinical Epidemiology, Jan 1995 48(1) pp61-66, 41 references

Outlines the new approach of EBM, comparing it with the traditional model which places a high value on authority, and discusses how it might be applied in practice. Focuses on NNT (the number of patients a clinician needs to treat with a particular therapy in order to prevent one clinical event) as a tool for therapeutic decision making and bedside teaching. For further general articles by David Sackett and William Rosenberg explaining the principles of EBM, and how it can be taught

See also: On the need for evidence-based medicine, by D L Sackett and W M Rosenberg (Health Economics, Jul/Aug 1995 4(4) pp249-54)

On the need for evidence-based medicine, by D L Sackett and W M Rosenberg (Journal of Public Health Medicine, Sep 1995 17(3) pp330-34, 22 references)

The need for evidence-based medicine, by D L Sackett and W M Rosenberg (Journal of the Royal Society of Medicine, Nov 1995 88(11) pp620-24, 22 references)

(32)

EVIDENCE-based medicine: what it is and what it isn't

Sackett, D L and others

British Medical Journal, 13 Jan 1996 312(7023) pp71-72, 14 references

Reviews the current interest in EBM whose philosophical origins are said to date back to the mid-nineteenth century and earlier. Summarises what is meant by EBM, and rejects some of the more frequently expressed criticisms of it. Notes the possible hi-jacking of

EBM by those wishing to cut health care costs, arguing that this suggests a fundamental misunderstanding of the financial consequences of this approach to medicine. For letters in response to this editorial which suggest that David Sackett's more moderate tone has not convinced everyone

See also: Evidence based medicine [letters] (British Medical Journal, 20 Jul 1996 313 (7050) pp169-71)

(33)
EVIDENCE-based medicine: Socratic dissent
Grahame-Smith, D
British Medical Journal, 29 Apr 1995 310(6987) pp1126-27
A dialogue between Socrates and Enthusiasticus which briefly explores many of the contradictions and problems inherent in EBM. Socrates concludes that the recent espousal of EBM by politicians and health service managers who 'see your beloved evidence based medicine as a means to shackle the doctors and bend them to their will' is a very dangerous development.

2. EVIDENCE-BASED MEDICINE AS A POLICY ISSUE

2.1 Government interest

Given the practical difficulties involved in introducing EBM, and the scepticism of conservatives in the medical profession, it is not surprising that in 1979 Archie Cochrane was still bemoaning the fact that his profession had failed to organise 'a critical summary, by specialty or subspecialty, of all relevant randomised controlled trials'. Four years later the Griffiths Report on the management of the NHS remarked that the 'clinical evaluation of particular practices is by no means common, and the economic evaluation of those practices [is] extremely rare'[34]. The latter half of this sentence is particularly instructive.

The rise of EBM as a purely professional response to the perceived need for improvements in the quality of care for individual patients has coincided with the increasingly urgent need to find political and managerial solutions to the spiralling costs of a public service for which there is infinite demand. The NHS reforms of the 1990s were prompted by a major financial crisis in 1987, and by the growing pressures of a number of fundamental trends in health service provision which are affecting all advanced economies[35]:

- The increasing mis-match between the growth in medical scientific knowledge and the ability of traditional mechanisms to maintain the knowledge and skill of practitioners. This requires a move away from the simple acquisition of facts towards the development of information seeking, problem solving and decision making skills.
- The increasing organisational complexity and specialisation of health care delivery in which individual patients are cared for by several practitioners, departments and organisations which may communicate poorly. Proper coordination of care is a major challenge.
- Increasing expectations of health care, resulting in ever closer scrutiny by the public and policy makers. Health services must operate effectively and efficiently, and be able to demonstrate the fact in terms of improved outcomes.
- Increasing health care costs as a result of medical advance and rising demand prompted by demographic factors. Managers need to contain and control costs while ensuring equity and high quality services.

A managerial tool

EBM, with its emphasis on eradicating useless or wasteful practices and concentrating on those which deliver the best outcome has enormous attractions as a tool for targeting scarce resources, and is fully in line with the commitment of both major political parties to the promotion of efficiency and monitoring of performance. In a situation in which resources are limited by definition, what better way of allocating them than on the basis of the best available research evidence as to therapeutic or diagnostic effectiveness, backed up wherever possible by information on cost-effectiveness? All restrictions on health care expenditure are likely to be unpopular with the electorate but at least evidence-based restrictions can be portrayed as objective and fair, even if unfortunate.

EBM has equal attractions for politicians and managers as an internal performance assessment tool to ensure value for money in the use of public funds and promote the dissemination of best practice. Thus the evidence-based process of clinical audit, introduced by the 1989 White Paper *Working for patients*[36], is now widespread within the NHS and involves doctors, nurses and many other health care professionals in the systematic review of the care and treatment they offer patients. As a result of (or as part of) the audit process, evidence-based clinical guidelines have been developed in many areas to underpin good practice.

The theme of concentrating resources on key priorities, identifying the most effective interventions and ensuring that they are translated into practice pervades all the major health policy documents of the 1980s and 1990s from both government and Parliament. The July 1992 White Paper on *The health of the nation*[37] identified five key areas on which effort and resources should be focused (coronary heart disease and stroke, cancers, mental illness, accidents, and HIV/AIDS and sexual health) and set targets for improvement. It states that 'the overall purpose of the NHS is to secure, through the resources available, the greatest possible improvement in the physical and mental health of the people of England'. This requires the health service to:

- Increase the knowledge base of clinical and cost-effectiveness.
- Ensure that this knowledge reaches decision makers.
- Encourage its use to change practice.
- Monitor results.

Improved information is an important element in the strategy, and the White Paper was followed by the publication of specifications for indicators to be used in monitoring progress towards the achievement of Key Area targets. In 1993 the first edition of a new annual *Health survey for England* was published[38] and the NHS has developed a wide range of monitoring activities to provide information of use to clinicians and health service managers.

The House of Commons Health Committee also addressed priority-setting issues in two reports in the mid-1990s dealing with the drugs budget and purchasing[39,40]. In the second, the Committee expressed considerable concern at the possibility that 'some routine treatments may, for most patients, be ineffective and a waste of resources', and clearly saw improved effectiveness information as an important way of releasing resources for other more beneficial uses. It warmly welcomed the establishment of the NHS Centre for Reviews and Dissemination, the UK Cochrane Centre and other initiatives to improve the dissemination and use of information, and urged the NHS Executive to ensure that 'all health commissions establish local arrangements actively to promote and distribute among local decision makers relevant information on effectiveness'.

Role of R&D

A key role in the campaign for a more effective health service is played by the NHS R&D strategy, developed following an influential report in 1988 by the House of Lords Select Committee on Science and Technology[41] which argued strongly that 'the NHS should articulate its needs [for research]; it should assist in meeting those needs; and it should ensure that the fruits of research are systematically transferred into service'. It spelt out the potential role of the Department of Health in promoting effective R&D and led directly to the appointment of the first NHS Director of R&D and the publication of the R&D strategy in 1991. As discussed in the next section, this has been at the forefront of the NHS Executive's drive to create an evidence-based health service.

Its emphasis on identifying key research priorities in consultation with the scientific community, and those working in and using, the health service is fully in tune with broader UK science policy as laid down in the May 1993 White Paper, *Realising our potential*[42] and the Technology Foresight Programme which has followed it. However, the NHS has arguably done a great deal more to emphasise the crucial importance of disseminating research findings and making sure that they are translated into practice. A report from the Advisory Council on Science and Technology, also in 1993[43], summarised it thus: 'currently the introduction of new technologies in health care in the UK can be a haphazard process', and a key challenge will be to ensure 'active take-up and widespread use of...validated procedures'.

Few EBM advocates within the medical profession would quarrel with the need for the prudent use of limited resources, for targeting key priorities, and for the regular monitoring of performance. Nor would they necessarily disagree that a science-based rationalisation of health care has the potential to release substantial resources, estimated by Sir Michael Peckham when he left his post as NHS director of R&D in early 1996, to be at least £1 billion. Some optimists might even agree with Sir Michael that EBM could be the key to a financially manageable NHS in which growing consumer demand can be accommodated without unacceptable pressure on the nation's finances.

A smokescreen for rationing?

However, there are those who are worried by the possibility that the original purpose of EBM to improve the effectiveness and quality of care for the individual patient – a process which may just as easily lead to an increase in care costs as to a decrease – is being obscured by the intensity of interest in its potential as a cost cutting or containment tool. David Hunter of the Nuffield Institute of Health in Leeds warns that EBM may be 'overly rooted in a narrow scientism' that values only those things which can be measured and quantified, and risks 'being oversold as a panacea for the problems of rationing and alleged underfunding in the NHS'[44]. The problem lies in the fact that it will take several years for EBM to have a significant and permanent impact, while the horizons of health service policy makers are much shorter term.

David Hunter doubts whether EBM, with its 'immaturity' and 'bias towards a narrow scientism' can really deliver the kind of improvements in medical care that its proponents claim, and certainly not within the time horizons of policy makers. As a result, there is a danger that the latter will lose interest in its 'inflated claims to redirect the course of modern medicine', and focus even more closely on its potential as a straightforward managerial tool. Those who seek to expand the use of EBM beyond the horizon of the individual clinician caring for the individual patient are well aware of the possibility that its principles may be distorted once it is absorbed into the contracting process. Thus an article on evidence-based purchasing in the May/June 1996 issue of *Evidence-Based Medicine* warns of the danger that 'purchasers may use evidence-based purchasing to focus on reducing the provision of services thought to be ineffective while paying no attention to increasing those thought to be effective, which would turn evidence-based purchasing into a tool for cost minimisation rather than value maximisation'[45].

In the last year or so, Labour Party spokesmen have regularly accused the Conservative government of using EBM as a smokescreen for the introduction of rationing, an emotive word which is routinely used as a term of abuse in the health care debate. Although it is increasingly accepted that rationing is inevitable in health care (and probably always has been), it is still felt by many politicians that to admit the fact publicly is tantamount to political suicide. Thus those in power must find acceptable euphemisms to describe hard decisions, while those in opposition are free to accuse them of deceit.

The November 1996 White Paper, *A service with ambition,* tackled head-on the claims of a variety of reputable authorities (including former NHS Chief Executive, Sir Duncan Nichol) that the health service was becoming unaffordable, and ruled out explicit rationing in a strong endorsement of the founding ideals of the NHS[46]. There should, it said, be 'no clinically effective treatments which a health authority decides as a matter of principle should never be provided'. The White Paper nevertheless acknowledged that 'difficult choices' will have to be made about priorities because of limited resources, and noted the contribution being made by the process of identifying and adopting those technologies which confer real benefit, while discouraging less clinically and cost-effective interventions. However, there was no suggestion of a national evidence-based list of preferred treatments: 'no such list could ever hope to accommodate the range and complexity of the different cases which clinicians face all the time', and there may be instances in which a treatment judged to be less effective in general terms will be appropriate for individual patients.

The White Paper may have specifically excluded rationing 'as a matter of principle' but this does not necessarily rule it out as a matter of financial necessity. Nor are cleverly worded attempts to avoid an outright admission of rationing solely the preserve of Conservative governments. It is highly unlikely that the incoming Labour government will have the scope to provide substantial new resources for the NHS, if only because of its manifesto commitment to stick to Conservative departmental spending ceilings for the next two years. Moreover, its ability to make large savings through modification of the internal market reforms and reductions in bureaucracy may well be limited. Labour health secretaries, just as much as their Conservative predecessors, will have to find acceptable ways of explaining 'difficult choices', and clearly see EBM as a useful tool in this context[47].

There may be signs, however, that the future health debate will be rather more honest than it is at present. It is increasingly difficult for the lay observer to see the distinction between 'difficult choices' and rationing, and elements within the medical profession are clear that it is time to come clean and admit that there have always been instances in which patients have been denied effective treatments because of resource constraints. The Rationing Agenda Group – which includes health academics, doctors, health service managers and members of health policy think tanks – argues that the way ahead is to involve the public in the debate and to abandon the pretence that the NHS can (or should) provide all possible treatments, whatever the cost and however slim the chance of a successful outcome[48].

Policy makers may be understandably cautious about stepping into this particular minefield given the volatility of public attitudes on health matters. While many ordinary people may accept the inevitability of rationing when they are in good health, their views on the matter are likely to be quite different if they or their loved ones face serious illness. Powerful emotions can also be raised by high profile cases like that of Jaymee Bowen whose robust approach to her disease led many to argue that the struggle to preserve life is more important than any financial – even clinical – consideration. These are difficult issues for doctors, health managers and politicians to deal with, and it is clear that the principles of EBM will continue to be crucially important both in making hard decisions on treatment options and resource allocation, and in explaining them to the public. As such, they will be indelibly associated with the rationing debate however much the EBM community might wish otherwise.

(34)
NHS management inquiry: Griffiths NHS Management Inquiry report
Department of Health and Social Security: London, 1983. 25pp

(35)
HEALTH care costs: an international dilemma
Abel-Smith, B
LSE Magazine, Summer 1996 8(1) pp7-9
An analysis based on a chapter to which Brian Abel-Smith contributed in *Choices in health policy: an agenda for the European Union* (Dartmouth, and Office for Official Publications, 1995). Looks at the inexorable rise of health care costs and at the strategies which governments are having to adopt to contain expenditure. EBM approaches are a key element, but are both costly and complex to implement. For a recent American analysis of the problems, whose title indicates the financial imperative behind health technology assessment
See also: Medicine worth paying for: assessing medical innovations, edited by H S Frazier and F Mosteller (Harvard University Press: Cambridge, MA, Nov 1995. 336pp)

(36)
WORKING for patients
Department of Health
HMSO (now Stationery Office), 1989. 101pp (Cmnd 555)

(37)
THE HEALTH of the nation: a strategy for health in England
Department of Health
HMSO (now Stationery Office), Jul 1992. 128pp (Cm 1986) (36pp summary also published)
Sets out a strategy for health in England, taking account of the response to consultation. Identifies five key areas for action: coronary heart disease and stroke, cancers, mental illness, accidents, and HIV/AIDS and sexual health. Indicates the action needed to achieve those targets, and outlines a framework for monitoring, development and review of the strategy. Annual progress reports have been published in following years. For details of the indicators to be used in monitoring progress towards the targets, and for the Department of Health's advice on each of the key areas
See also: Specification for national indicators (Department of Health: London, 1992. 128pp)
Accidents (Department of Health: London, Jan 1993. 99pp. Key Area Handbook)
Cancers (Department of Health: London, Jan 1993. 169pp. Key Area Handbook)
Coronary heart disease and stroke (Department of Health: London, Jan 1993. 115pp. Key Area Handbook)
HIV/AIDS and sexual health (Department of Health: London, Jan 1993. 199pp. Key Area Handbook)
Mental illness (Department of Health: London, Jan 1993. 163pp. Key Area Handbook)

(38)
HEALTH survey for England 1991
White, A; Nicolaas, G
HMSO (now Stationery Office), 1993. 373pp (HS 1)
The first in an annual series of surveys designed to monitor trends in the nation's health.
Includes basic descriptive statistics for the variables measured, and information on their
associations with demographic, social and behavioural factors.

(39)
**PRIORITY setting in the NHS: the NHS drugs budget: 2nd report, session
1993-94**
House of Commons Health Committee
HMSO (now Stationery Office), 1994. 2 vols (HC 1993-94 80-I/II)
Looks at the increase in expenditure on drugs, driven primarily by technological advance,
changes in therapeutic approaches and demographic factors. As a largely natural
phenomenon, it should not be regarded as a major cause for alarm, although attention
should be paid to ways of making savings wherever possible. Encouraging more rational
and effective prescribing is the most promising approach, and the Committee recommends
a National Prescribing List. For the government's response
See also: Government response to the 2nd report from the Health Committee, session
1993-94: priority setting in the NHS: the NHS drugs budget, by Department of Health
(HMSO, 1994. 24pp. Cm 2683)

(40)
PRIORITY setting in the NHS: purchasing: first report, session 1994-95
House of Commons Health Committee
HMSO (now Stationery Office), Feb 1995. 2 vols (HC 1994-95 134-I/II)
Admits that there is no such thing as a correct or ideal set of priorities, but argues that
greater consistency is needed across the health service. The use of clinical and cost
effectiveness data will be a major factor in this context, although coercion of service
providers is to be avoided. Persuasion is the key to changing behaviour. For the
government's response
See also: Government response to the 1st report from the Health Committee, session
1994-95; priority setting in the NHS: purchasing, by Department of Health (HMSO, Apr
1995. 18pp. Cm 2826)

(41)
PRIORITIES in medical research: 3rd report, session 1987-88
House of Lords Select Committee on Science and Technology
HMSO (now Stationery Office), 1988. 3 volumes (68pp, 408pp, 390pp) (HL 54 I/II/III)
This influential report spelt out the potential role of the Department of Health in
promoting effective R&D, and led to the creation of the R&D Directorate and the NHS
R&D strategy which was published in 1991. For the government's response
See also: Priorities in medical research: government response to the 3rd report of the
House of Lords Select Committee on Science and Technology, by Department of Health
(HMSO, 1989. 11pp. Cm 902)

(42)

REALISING our potential: a strategy for science, engineering and technology

Office of Science and Technology

HMSO (now Stationery Office), May 1993. 74pp (Cm 2250)

This White Paper underpins UK science policy in all fields, and aims to link the country's research effort more closely to wealth creation and improved quality of life. It put in place the Technology Foresight Programme which is designed to identify future research needs and priorities, and to guide the country's research effort. For the report of the foresight analysis of health-related research

See also: Health and life sciences (HMSO, Mar 1995. 112pp. Progress Through Partnership 4)

(43)

A REPORT on medical research and health

Advisory Council on Science and Technology

HMSO (now Stationery Office), 1993. 48pp

Report of an inquiry set up in parallel with the Peckham review of the NHS R&D strategy.

See also: The government's response to the ACOST report on medical research and health (Department of Health: London, Jun 1993. 23pp)

(44)

RATIONING and evidence-based medicine

Hunter, D J

Journal of Evaluation in Clinical Practice, Feb 1996 2(1) pp5-8, 11 references

Discusses the limitations of EBM which is both immature and based on an untested belief in scientific positivism. As such, it makes a spurious claim to provide certainty in a world of uncertainty. Suggests that EBM may fail in its claims to redirect the development of modern medicine, and may be taken over by health service managers as an explicit tool of rationing to give a scientific gloss to decisions which are really value judgements that should be widely discussed with the public. For more from David Hunter on the rationing debate, and for further comment on the inherent uncertainty of medicine and the dangers of scientism

See also: Rationing dilemmas in health care, by D J Hunter (National Association of Health Authorities and Trusts (now NHS Confederation), Birmingham Research Park, Vincent Drive, Birmingham B15 2SQ, 1993. 36pp. Research Paper 8)

Rationing health care, by S Harrison and D J Hunter (Institute for Public Policy Research, 30-32 Southampton Street, London WC2E 7RA, 1994. 89pp)

Evidence-based medicine: the illusion of certainty, by D J Hunter (IFMH Inform, Autumn/Winter 1996 7(3) pp1-3, 6 references)

Scientism and economism in the regulation of health care, by D M Frankford (Journal of Health Politics, Policy and Law, 1994 19(4) pp773-99)

The NHS and the new scientism, by R Klein (Quarterly Journal of Medicine, Jan 1996 89(1) pp85-87)

(45)

EVIDENCE-BASED purchasing

Milne, R; Hicks, N

Evidence-Based Medicine, May/Jun 1996 1(4) pp101-02, 13 references

Discusses the potential of EBM in the wider context beyond the individual patient, looking at what is needed to make evidence-based purchasing a reality. Also identifies potential dangers including the possibility that purchasers may use EBM as a cost minimisation tool. For more brief comment on evidence-based purchasing

See also: The informers, by J Farmer and R Chesson (Health Service Journal, 1 Feb 1996 pp28-29)

(46)
THE NATIONAL Health Service: a service with ambition
Department of Health
Stationery Office, Nov 1996. 60pp (Cm 3425)
Argues that health care should remain free and universally available, and promises guaranteed, long term, real funding for growth. Rules out explicit rationing and the imposition of a national list of preferred treatments of proven efficacy, while admitting that 'difficult choices' need to be made.

(47)
RENEWING the NHS: Labour's agenda for a healthier Britain
Labour Party, 150 Walworth Road, London SE17 1TJ, 1995. 39pp

(48)
THE RATIONING agenda in the NHS
New, W
British Medical Journal, 22 Jun 1996 312(7046) pp1593-1601, 6 references
A spokesman for the Rationing Agenda Group presents, as neutrally as possible, the various issues relating to rationing and priority setting in the NHS. Concentrates on the delivery rather than financing of health care, and includes comment on specific cases such as that of Jaymee Bowen. For more recent comment on the rationing (or priority-setting) debate

See also: Rationing revisited: a discussion paper, by British Medical Association, Health Policy and Economic Research Unit (British Medical Association, BMA House, Tavistock Square, London WC1H 9JP, 1995. 16pp. Discussion Paper 4)
Setting priorities in the NHS: a framework for decision-making (Royal College of Physicians, 11 St Andrews Place, London NW1 4LE, 1995. 42pp)
Is health care rationing really necessary? by P Mullen (University of Birmingham, Health Services Management Centre, 40 Edgbaston Park Road, Birmingham B15 2RT, Nov 1995. 34pp. Discussion Paper 36)
Hard choices: priority setting in the NHS, by W Moore (National Association of Health Authorities and Trusts (now NHS Confederation), Birmingham Research Park, Vincent Drive, Birmingham B15 2SQ, 1996. 20pp)
Rationing and rights in health care, by J Lenaghan (Institute for Public Policy Research, 30-32 Southampton Street, London WC2E 7RA, 1996. 102pp)
Rationing in the NHS: principles and pragmatism, by J New and J Le Grand (King's Fund, 11-13 Cavendish Square, London W1M 0AN, 1996. 82pp)

2.2 The National Health Service response

A research and development strategy

Although the beginnings of NHS interest in evidence-based practice can be seen in the development of clinical audit, the main starting point was the arrival of Sir Michael

Peckham as the first Director of R&D for the Department of Health and the NHS Executive in April 1991, and the subsequent publication of the health service R&D strategy in September of the same year[49]. The aim of the strategy was to create an 'evaluative culture' in a service in which 'there is a generally poor uptake of research findings by clinicians and health service managers'. It outlined plans to switch most of the health service R&D budget into the assessment of new drugs, medical equipment and other treatments, and introduced a number of institutional and managerial innovations:

- A Central Research and Development Committee (CRDC) to set the strategic R&D framework. The CRDC is served by a range of advisory groups which examine R&D needs in specific areas. Advisory reviews have already been completed in many clinical areas such as mental health and learning disabilities, cardiovascular disease and stroke, physical and complex disabilities, cancer, mother and child health, and asthma management. In addition, advisory groups have considered background issues such as the primary/secondary care interface, the provision of primary care data and the crucial EBM issue of methods of implementing research findings. Once advisory groups have completed their work they are disbanded and Commissioning Groups are established to take forward a programme of work in the priority areas.

- A regional R&D infrastructure in which each region has a Director of R&D and an R&D manager, with support staff, to manage nationally and regionally funded R&D programmes. Specific guidance to the regions was issued alongside the national R&D strategy, and complementary strategies were subsequently published for Wales and Scotland[50,51].

- A Standing Group on Health Technology to manage a central Health Technology Assessment Programme. A preliminary report on health technology assessment was issued in 1992[52], and the Group itself was established at the beginning of 1993. Like the advisory groups which have examined specific disease areas, it serves the CRDC but has permanent status. It looks at the use of devices, equipment, drugs and procedures across medical, nursing and other health care practices, and has commissioned over 50 studies to date, focusing on 'areas of uncertainty in the service'[53]. The Health Technology Assessment Programme is a cyclical process which identifies problem areas within the NHS, translates those problems into research questions, prioritises research topics, funds projects within the priority areas, disseminates research results and evaluates the impact on practice. It will spend some £20 million of NHS funding between 1995 and 2000.

- Information dissemination systems to promote the effective use and exchange of existing research information produced both inside and outside the NHS[54]. These include the UK Cochrane Centre established in October 1992; the NHS Centre for Reviews and Dissemination set up at the end of 1993; and the NHS R&D Project Registers System. There is a close relationship between such bodies and the Health Technology Assessment Programme because of the latter's need to establish the current status of research in 'areas of uncertainty', and to ensure that any NHS-funded research effort does not duplicate work already being carried out elsewhere.

In December 1993 the NHS Executive's broader commitment to an evidence-based health service was made clear in the first of a series of Executive Letters designed to increase the use of information on clinical and cost effectiveness. EL(93) 115 listed sources of information, commented on a limited number of high quality clinical guidelines, and gave information on interventions and services that were under investigation[55]. Crucially, it also asked purchasing authorities to provide evidence of the action they were taking in response to guidelines issued in connection with the 1994-95 contracting round, confirming that the NHS Executive's promotion of clinical effectiveness and efficiency was not simply to be a matter of information and encouragement.

In 1994 the Department of Health published its two volume *Health care needs assessments* directed primarily at public health physicians in purchasing authorities[56]. Covering 20 conditions and services which account for over a third of the 'burden of disease' in western countries, these epidemiologically-based assessments examined the range and effectiveness of services currently available in the NHS and made suggestions for change. Further Executive Letters in 1994 and 1995 pointed health service managers in the direction of yet more useful sources of information, but also confirmed that health authorities will have to demonstrate that they are taking clinical effectiveness seriously. As part of their performance management role, NHS regional offices will now review authorities' investment in promoting clinically cost effective practice on a regular basis.

Culyer Report

Despite these initiatives, there was still concern during the early 1990s that the NHS R&D effort needed further reform to back up the drive towards greater clinical effectiveness. Although the R&D strategy had ensured greater coherence, the funding for research still remained haphazard with continuing evidence of research duplication and a lack of targeting on priorities. In some cases, good quality research was failing to attract support while poor quality projects were allowed to go ahead. An NHS R&D Task Force was established under the chairmanship of health economist Professor Anthony Culyer, and reported in September 1994[57]. Its recommendations included:

- The establishment of a National Forum chaired by the NHS Director of R&D and including representatives from the NHS, health departments, research councils, the Association of Medical Research Charities, the Wellcome Trust, the higher education funding councils, the Association of the British Pharmaceutical Industry, the health care industries and the Office of Science and Technology. The Forum's role is to advise the Director of R&D and the Secretary of State on national and international developments relating to health service R&D, and help to ensure a coordinated approach across the whole spectrum of organisations involved in health-related research. In the latter context, Culyer also recommended that the 1992 NHS concordat with the Medical Research Council should be reviewed, and that the possibility of formal understandings with other external bodies should be explored.
- The expansion of the CRDC's role to include the provision of advice on how best to invest R&D funds, as well as setting the strategic R&D framework with the support of the National Forum. Membership was revised accordingly.
- The development of a stronger regional dimension for R&D. Although the strategy should continue to be set nationally, regional R&D directors should be the focal point for R&D effort.
- The introduction of 'single stream funding' for R&D, conceived as an annually determined levy on all health care purchasers' funding allocations. The aim is to rationalise a muddled funding system and protect R&D from the adverse effects of the internal market.

In December 1994, the Secretary of State for Health announced new measures which accepted Culyer's recommendations, and provided an additional £8 million for NHS-commissioned research in 1995-96[58]. In addition, new agreements were reached between the NHS and the Engineering and Physical Sciences Research Council, the

Biotechnology and Biological Sciences Research Council and the Economic and Social Research Council. These moves have helped to allay some fears about the impact of the NHS internal market reforms on curiosity-driven health-related research although, as a House of Lords report in June 1995 makes clear, there are still reservations about the impact of greater targeting on the excellence of the medical research base[59]. These mirror broader concerns in the scientific community over the potential damage to the science base of a science policy linked to explicit economic and social goals.

Just prior to the Secretary of State's announcement, the NHS Executive published details of how research and development was to fit into the NHS following a major Functions and Manpower Review[60]. Six key functions are envisaged for R&D, confirming yet again the Executive's commitment to an evaluative, evidence-based culture:

- Identifying requirements for new research.
- Ensuring that knowledge is forthcoming.
- Ensuring that information about research-based knowledge is available to decision makers.
- Promoting the use of information about research-based knowledge.
- Promoting an evaluative culture.
- Developing and evaluating the NHS R&D strategy.

Promoting clinical effectiveness

The drive for improved effectiveness has now become a central challenge for the NHS, pervading all aspects of clinical and related work and moving well beyond EBM's original focus on the quality of care provided by individual clinicians[61]. Health authorities and General Practice fundholders are expected to pursue cost, as well as clinical, effectiveness in a number of different, but complementary, ways. The role of EBM in identifying and applying best practice information from the enormous pool of medical research literature is certainly there, along with the role of the R&D strategy in monitoring health-related R&D and filling gaps. However, these are now paralleled by the assessment of cost effectiveness and clinical performance through audit and outcome assessment. It is no longer enough to know that a particular intervention exists and is capable of producing clinical benefits; an increasingly cost-conscious NHS also needs to know that it actually has produced those benefits, at acceptable cost, in particular situations.

The NHS Executive's advice to chief executives of health authorities and NHS Trusts on *Promoting clinical effectiveness* describes the approach as a three-sided triangle consisting of information, change and monitoring[62]:

- **Information** – this is provided by the R&D strategy, the Health Technology Assessment programme, and the work of national bodies such as the Cochrane Centre, the NHS Centre for Reviews and Dissemination and the Centre for Evidence-Based Medicine at Oxford. It also includes local initiatives such as the monthly *Bandolier* newsletter published by the Anglia and Oxford Region. All these are covered in more detail in Chapters 3 and 4.
- **Change** – this covers methods of ensuring that clinical effectiveness information gets into practice through initiatives such as clinical audit, clinical guidelines, education and training, and the provision of better information to patients. The NHS Executive is funding a multidisciplinary National Centre for Clinical Audit which will be overseen

by the Clinical Outcomes Group, while health authorities have been formally responsible for audit from April 1996 and will be regularly monitored by the Executive's regional offices. These and other change initiatives are discussed in Chapter 4, while the patient perspective is covered in Chapter 5.

- **Monitor** – some of the change initiatives, particularly clinical audit, will provide input to the third side of the triangle which also covers monitoring activities within the NHS Executive's Information Management and Technology Strategy. These include the development of a variety of national data sets such as the Public Health Common Data Set, population health outcome indicators in a range of areas, Health of the Nation NHS performance measures, and indicators based on primary care and prescription data. The NHS Executive is currently trying to draw together this bewildering array of monitoring information in a Compendium of Clinical Health and Outcome Indicators to be published in phases between June 1996 and mid-1997.

Clinical effectiveness is now formally included in NHS priorities and planning guidance as a medium term priority. Under the terms of the 1996-97 guidance, purchasers will have to demonstrate the sources of information they use to judge the effectiveness of services, the ways in which they use the results of clinical audit and outcome measurement to initiate change, and the ways in which they use information to give patients a better understanding of the effectiveness of treatments. The 1995-96 guidance required health authorities to show evidence of promoting at least two interventions known to be effective, and discouraging the use of at least two known to be less effective. In 1996-97 the requirement is more onerous[63].

Purchasers are expected to have developed strategies in association with providers and their clinicians, and primary health care teams, 'to secure sustained and comprehensive improvements in clinical effectiveness, which demonstrate: the use of evidence of clinical outcomes and the results of clinical audit to influence change in services; their sources and use of information to judge the effectiveness of services or interventions; [and] how patients are being informed about evidence of effectiveness related to their treatment.' As a result of these strategies 'purchasers should be able to demonstrate a significant change in the level of investment in an agreed range of primary, secondary and continuing care interventions' on the basis of evidence to show which interventions are more or less effective.

(49)
RESEARCH for health: a research and development strategy for the NHS
Department of Health, Richmond House, 79 Whitehall, London SW1A 2NS, Sep 1991. 15pp
Produced under the guidance of the NHS R&D Director, Sir Michael Peckham. For associated guidance to the regions, a later interview with Sir Michael, and his recent book on EBM written with Richard Smith
See also: NHS research and development strategy: guidance for regions (NHS Executive, Richmond House, 79 Whitehall, London SW1A 2NS, Sep 1991. 31pp)
Filling the lacuna between research and practice: an interview with Michael Peckham, by R Smith (British Medical Journal, 27 Nov 1993 307(6916) pp1403-07)
Scientific basis of health services, by M Peckham and R Smith (BMJ Publishing: London, 1996. 186pp)

(50)

SHARPENING the focus: a research and development framework for NHS Wales

NHS Directorate for Wales, Value For Money Unit, Cathays Park, Cardiff CF1 3NQ, Dec 1992. 28pp

The Welsh response to the Peckham Report. NHS Wales is to be a knowledge-driven, evaluation-based enterprise which will strengthen the scientific base of health care delivery, support the systematic development and application of research results in clinical practice, and encourage the growth of research-based health care industries in Wales. Also establishes an NHS Office for Research and Development, and appoints a director of R&D. For later Welsh thinking on clinical and cost effectiveness following the establishment in 1994 of a Welsh Clinical Effectiveness Group

See also: Towards evidence-based practice: a clinical effectiveness initiative for Wales (Welsh Office: Cardiff, and Central Office of Information, May 1995. 32pp)

Improving access to evidence and information: a clinical effectiveness initiative for Wales (Welsh Office: Cardiff, Aug 1995. 33pp. Briefing Paper 1)

Framework for the development of a multi-professional clinical audit in Wales: a clinical effectiveness initiative for Wales (Welsh Office: Cardiff, Feb 1996. 40pp. Briefing Paper 2)

Developing the working environment: a clinical effectiveness initiative for Wales (Welsh Office: Cardiff, Mar 1996. 24pp. Briefing Paper 3)

(51)

RESEARCH and development strategy for the National Health Service in Scotland

Scottish Home and Health Department

HMSO (now Stationery Office), 1993. 32pp

The Scottish response to the Peckham Report. The provision of clinical and cost effectiveness information has been led by the Clinical Resource and Audit Group, and the Joint Working Group on Purchasing which have launched a series of initiatives to disseminate information on effective clinical practice and ensure its adoption. They also aim to coordinate the research, dissemination and implementation activities of organisations across Scotland in an attempt to reduce duplication and encourage proper targeting of resources.

(52)

ASSESSING the effects of health technologies: principles, practice, proposals

Department of Health R&D Division, Richmond House, 79 Whitehall, London SW1A 2NS, 1992. 28pp

The report of a working group headed by Iain Chalmers which led subsequently to the establishment of the Health Technology Assessment programme. For a conference proceedings published during the same year, and for the views of the editor of the *British Medical Journal*

See also: Tidal wave: new technology, medicine and the NHS: the proceedings of the Caversham Conference on Health Technology Assessment sponsored by the Institute of Health Services Management and the Management Advisory Service to the NHS (King's Fund Centre, 11 13 Cavendish Square, London W1M 0AN, 1992. 34pp)

Towards a knowledge-based health service: priorities for health technology assessment, by R Smith (British Medical Journal, 23 Jul 1994 309(6949) pp217-18)

(53)

REPORT of the NHS Health Technology Assessment Programme 1995

Department of Health R&D Directorate, Quarry House, Quarry Hill, Leeds LS2 7UE, Oct 1995. 53pp

For the first report of the Standing Group on Health Technology

See also: Standing Group on Health Technology report (NHS Executive: Leeds, Jun 1994. 27pp)

(54)

THE NHS Information Systems Strategy

Ennis, J and others

Health Libraries Review, Dec 1994 11(4) pp227–34, 4 references

A useful summary of the Strategy which is designed to support R&D within the health service. Key elements include the UK Cochrane Centre, the NHS Centre for Reviews and Dissemination, and the NHS Project Registers System. For the study which preceded the establishment of the Strategy, and for an updated summary of what it involves

See also: An information management and technology strategy for the NHS in England: a view for commissioning agencies, by Information Management Group UK (NHS Executive, Richmond House, 79 Whitehall, London SW1A 2NS, 1992. 74pp)

The information management and technology strategy for the NHS in England: an updated overview (NHS Executive, Information Management Group, Feb 1996. 6pp)

(55)

IMPROVING clinical effectiveness

NHS Executive, Richmond House, 79 Whitehall, London SW1A 2NS, Dec 1993. (EL (93) 115)

The first Executive Letter to make explicit the NHS Executive's commitment to improving clinical effectiveness and moving to EBM. The recommended guidelines cover asthma management, the management of diabetes in primary care, the treatment of head injury, the investigation and management of stable angina, the management of leg ulcers in the community, the development of audit measures and guidelines for managing neonatal respiratory disease syndrome, and making the best use of a department of clinical radiology. For subsequent Executive letters on the effectiveness theme

See also: Improving the effectiveness of the NHS (1994. EL(94) 74)

Effectiveness: a vision for the future (1995. EL(95) 13)

Improving the effectiveness of clinical services (1995. EL(95) 105)

(56)

HEALTH care needs assessment

NHS Executive, Richmond House, 79 Whitehall, London SW1A 2NS, 1994. 2 volumes

The assessments cover alcohol misuse; cancer of the lung; cataract surgery; colorectal cancer; the community child health service; diabetes mellitus; dementia; drug abuse; family planning, abortion and fertility services; hernia repair; lower respiratory disease; mental illness; people with learning disabilities; prostatectomy for benign prostatic hyperplasia; renal disease; stroke; total hip replacement; total knee replacement; and varicose vein treatment.

(57)
SUPPORTING research and development in the NHS
NHS Research and Development Task Force
HMSO (now Stationery Office), Sep 1994. 90pp
The Culyer Report. For the NHS Executive's subsequent implementation plan, and for guidance on associated cost issues for the English regions and Wales
See also: Supporting research and development in the NHS: implementation plan: a plan for implementing the recommendations of the NHS R&D Task Force (NHS Executive, Richmond House, 79 Whitehall, London SW1A 2NS, Apr 1995. 87pp)
Supporting research and development in the NHS: a declaration of NHS activity and costs associated with research and development: initial guidance (Department of Health, Richmond House, 79 Whitehall, London SW1A 2NS, Sep 1995. 37pp)
Supporting research and development in the NHS in Wales: a declaration of NHS activity and costs associated with research and development: initial guidance (University of Wales College of Cardiff, Office of Research and Development for Health and Social Care, Dec 1995. 28pp)
Supporting research and development in the NHS in Wales: a declaration of NHS activity and costs associated with research and development: guidance on costing and making the declaration (University of Wales College of Cardiff, Office of Research and Development for Health and Social Care, Jan 1996. 22pp)

(58)
CENTRALLY commissioned research programme
Department of Health, Richmond House, 79 Whitehall, London SW1A 2NS, Jan 1995. 71pp

(59)
MEDICAL research and the NHS reforms: 3rd report, session 1994-95
House of Lords Select Committee on Science and Technology, Sub-committee 1
HMSO (now Stationery Office), Jun 1995. 2 volumes (House of Lords Paper 12-I/II)
Argues that the excellence of the UK's medical research is threatened by the government's market-oriented NHS reforms, with damage already evident in academic medicine and curiosity-driven research. However, if the government implements the recommendations of the Culyer Report, the NHS regional R&D directors should have sufficient resources to support an effective research programme. For the government's response
See also: Medical research and the NHS reforms: government response to the 3rd report of the House of Lords Select Committee on Science and Technology: 1994-95 session, by Department of Health and others (HMSO, Sep 1995. 20pp. Cm 2984)

(60)
RESEARCH and development in the new NHS: functions and responsibilities
NHS Executive, Richmond House, 79 Whitehall, London SW1A 2NS, Nov 1994.
For the report on the Functions and Manpower Review which forms the background for this document
See also: Managing the new NHS: functions and responsibilities in the new NHS (Department of Health: London, and NHS Executive, 1994. 48pp)

(61)
RESEARCH and development: towards an evidence-based health service
Department of Health, Research and Development Division, Richmond House, 79
Whitehall, London SW1A 2NS , Jun 1995. Looseleaf (59pp)
Outlines the development of NHS policy on this issue, with comment on the 1991 R&D
strategy and the subsequent Culyer Report. Includes a wide range of information covering
areas such as the National Forum; the Central Research and Development Committee;
priority setting; the coordination and commissioning of new research; current NHS R&D
programmes; the information systems strategy; relationships between the NHS R&D
effort and both national and European research policies and programmes; concordats with
the research councils; and R&D links with other external bodies. For a recent analysis of
R&D in the NHS
See also: Research and development for the NHS: evidence, evaluation and effectiveness,
edited by M Baker and S Kirk (Radcliffe Medical Press: Oxford, 1996. 125pp)
Note: information on the progress and development of the NHS R&D strategy is also
available on the Internet through a variety of routes including the Department of Health's
R&D Strategy Home Page at **http://www.open.gov.uk/doh/rdd1.htm**

(62)
**PROMOTING clinical effectiveness: a framework for action in and through the
NHS**
NHS Executive, Richmond House, 79 Whitehall, London SW1A 2NS, Jan 1996. 47pp
For a summary of the NHS 'triangle' approach from the Medical Director of the NHS
Executive, and for a *BMJ* editorial which comments on the limitations of the Executive's
approach and hopes that the Department of Health's policy activities will also be subject to
evidence-based scrutiny
See also: Improving clinical effectiveness: a co-ordinated approach, by G Winyard (In:
Clinical effectiveness: from guidelines to cost-effective practice, edited by M Deighan and
S Hitch. Earlybrave Publications Limited, PO Box 3165, Brentwood, Essex CM13 1TL,
Oct 1995 pp1-6)
Promoting clinical effectiveness, by J Hayward (British Medical Journal, 15 Jun 1996 312
(7045) pp1491-92, 19 references)

(63)
PRIORITIES and planning guidance 1996/97
NHS Executive, Richmond House, 79 Whitehall, London SW1A 2NS, 1995. 21pp (EL
(95) 68)
The 1995/96 guidance was published in 1994 as EL(94) 55.

3. SOURCES OF EVIDENCE

The first side of the NHS Executive's clinical effectiveness triangle consists of information. The number of sources potentially available to a doctor, other health care professional or manager is bewildering, ranging from highly specialised refereed clinical journals reporting on new research to systematic reviews of published research and specific analyses of the clinical or cost effectiveness of particular interventions. Some are national or international in origin, others are produced at the regional or local level. Many different organisations are involved including:

- The NHS itself at both national and regional level. Initiatives include the compilation of meta-analyses and reviews, as well as the development of audit and outcome measurement methodologies, and clinical guidelines.
- Academic units such as the Centre for Health Economics at the University of York, the Health Services Management Centre at the University of Birmingham, the University of Bristol's Health Care Evaluation Unit, and the Brunel Health Economics Research Group. These bodies are engaged in direct evaluative work in health care or health economics, and also contribute to the EBM process through their teaching and research.
- Independent research organisations, and pharmaceutical companies and other suppliers of health care products. Information is also available in a variety of different media for different audiences – technical journals in the clinical and health economics fields; newsletters; briefing series; conference reports; online databases; CD-ROMs; and, increasingly, the Internet[64].

In addition to the output of these bodies, there is also a substantial and growing secondary literature in medical and other journals on the processes involved in identifying relevant studies for systematic reviews, assessing them, and creating and managing the subsequent output. This deals with a wide range of issues including the quality of primary research studies, and reviews of them; the inadequacies of abstracts in clinical journals; the deficiencies of bibliographic database indexing; the minutiae of manual journal searching; controlling for publication bias or the under-reporting of research in the compilation of systematic reviews, often associated with calls for the development of an international registry of clinical trials; and making sure that reviews are kept up to date. There is also a significant literature on the subjects of clinical audit and clinical guidelines.

This chapter makes no attempt at comprehensive coverage of what is a confusing, fragmented and constantly evolving picture, but focuses on some of the major institutional sources which can be used by those wishing to identify primary EBM information. Others concentrating particularly on EBM teaching and learning are included in Chapter 4, together with some of the regional NHS initiatives. In addition, useful short cuts to finding out about EBM sources on the Internet are provided by:

- **OMNI (ORGANISING MEDICAL NETWORKED INFORMATION) PROJECT**
  The Library
  National Institute for Medical Research

Mill Hill

London NW7 1AA

Email: help@omni.ac.uk/ (for comments, questions and suggestions)

Email: omni:mailbase.ack.uk (to join the OMNI list send a message saying join omni your-first-name your-last-name)

URL: **http://www.omni.ac.uk**

A gateway to Internet resources in medicine, biosciences, allied health and health management, providing comprehensive coverage of UK resources in this area and access to other good quality resources world-wide. The OMNI service, which was launched in November 1995, is funded by the Joint Information Systems Committee of the Higher Education Funding Councils and is managed as part of the Electronic Libraries Programme (eLib). The lead body is the National Institute for Medical Research Library, with other consortium members including the Wellcome Centre for Medical Science and the medical libraries of Nottingham University, Cambridge University, the Royal Free Hospital School of Medicine, and the King Edward Hospital Fund.

All resources entered into OMNI are assessed for quality and regularly reviewed on the basis of detailed guidelines for selection and evaluation. A booklet, *Internet resources for medicine and bioscience: 2nd edition*, is available giving details of 'exemplary sites' which demonstrate the kinds of resources accessible through OMNI including educational/teaching materials; factual databases; bibliographic databases; directories; electronic journals, books and other full text services; software collections; and discussion groups. A monthly newsletter has been published since April 1996 (available via email and the OMNI server at the URL: **http://www.omni.ac.uk/general-info/newsletter.html**) and the 1996 annual report gives details of progress (and problems) during the first year of operation. It notes that accesses to the OMNI Welcome page are running at some 3,400 per week.

- **SHEFFIELD CENTRE FOR HEALTH AND RELATED RESEARCH**
  Regent Court
  30 Regent Street
  Sheffield S1 4DA

  Tel: 0114 276 8555 ext. 5454/5455
  Fax: 0114 272 4095
  Email: scharrlib@sheffield.ac.uk
  URL: **http://www.shef.ac.uk/uni/academic/R-Z/scharr/ir/netting.html**

  Andrew Booth at SCHARR produces a regularly updated introduction to evidence-based practice on the Internet called Netting the Evidence which lists a range of EBM sources and products; and a hard copy bibliography and resource guide entitled *The SCHARR guide to EBP (evidence-based practice)* (Apr 1996. 59pp. Occasional Paper 2). He has also contributed a recent article to *Health Libraries Review* on EBM issues[65].

- **UNIVERSITY OF HERTFORDSHIRE LIBRARY**
  Library and Media Services
  College Lane
  Hatfield
  Hertfordshire AL10 9AD

Tel: 01707 284678
Fax: 01707 284670
Email: c.cox@herts.ac.uk
URL: **http://www.herts.ac.uk/lrc/subjects/health/**

Library and Media Services provide central learning resources for the University at its various campuses which all have service points. The health information site maintained by Chris Cox includes a section on EBM which comprises definitions of EBM; an up-to-date listing of journal articles and other published material on EBM; news items and details of forthcoming conferences; and access to a range of EBM Internet resources.

In addition, a number of organisations have set up listings of, or pages of links to, other EBM resources. For example:

- Cambridge University Public Health: **http://fester.his.path.cam.ac.uk/phealth/phweb.html**
- Centre for Evidence-Based Medicine: **http://cebm.jr2.ox.ac.uk/docs/otherebmgen.html**
- Health Promotion Research Internet Network: **http://www.dsg.ki.se/socmed**
- McMaster University: **http://hiru.mcmaster.ca/ebm/**
- NHS Centre for Reviews and Dissemination: **http://www.york.ac.uk/inst/crd/welcome.htm**
- Southampton University: **http://www.soton.ac.uk/~swhclu/ebm.htm**
- University of Nottingham: **http://www.ccc.nottingham.ac.uk/~mpzjlowe/evpath.html**
- Oxford Clinical Information: **http://www.health.su.oz.au/resource/netting.htm** and **http://panizzi.shef.ac.uk/auracle/link2.html**

(64)
EVIDENCE-BASED medicine, the Internet, and the rise of medical informatics
Coiera, E
Hewlett-Packard Laboratories: Bristol, Feb 1996. 7pp, 14 references (HPL-96-26)
Looks briefly at the emergence of medical informatics, the move towards EBM and the rise of the Internet. Argues that the Internet can provide evidence-based practice with some of the key tools it needs to achieve its aims. However, it has its drawbacks, in particular by permitting the uncontrolled publication of scientific research without peer review. For more from Enrico Coiera, and for a general analysis of medical resources on the Internet
See also: Recent advances in medical informatics, by E Coeira (Hewlett-Packard Laboratories, Apr 1995. 19pp. HPL-95-41)
Medical informatics, by E Coeira (British Medical Journal, 27 May 1995 310(6991) pp1381-87)
Medical resources on the Internet, by J V Glowniak (Annals of Internal Medicine, 15 Jul 1995 123(2) pp123-31 + commentary, pp152-53)

(65)
FROM intelligence to evidence-based healthcare: a purchasing odyssey
Booth, A; Hey, S
Health Libraries Review, Dec 1995 12(4) pp249-59
Looks at the development of 'purchasing intelligence facilities', staffed in the main by
health librarians, and discusses the importance of the concept of intelligence to the NHS as
a whole.

3.1 Centre for Evidence-Based Medicine

Level 5
John Radcliffe Hospital
Headley Way
Headington
Oxford OX3 9DU

Tel: 01865 221320
Fax: 01865 222901
Email: david.sackett@ndm.ox.ac.uk
URL: **http://cebm.jr2.ox.ac.uk**

The Centre for Evidence-Based Medicine (Director, David Sackett) was set up in 1994 as
a joint venture involving the Anglia and Oxford Regional Health Authority, the NHS
R&D programme, the Nuffield Department of Medicine at the University of Oxford, the
Oxford-Radcliffe Hospitals Trust, and the Oxford Institute of Health Sciences. It is
expected to be the first of a series of such centres, and opened formally in March 1995 with
the twin aims of:

- Promoting the teaching, learning, practice and evaluation of EBM and evidence-based
  health care. As outlined briefly in Chapter 4, the Centre's educational initiatives include
  running workshops in EBM skills and the teaching of EBM, and collaborating with
  other institutions in the development of postgraduate training in the conduct of RCTs
  and systematic reviews. In addition, Centre members act as EBM pioneers by following
  its principles in their day-to-day clinical work.
- Conducting applied, patient-based and methodological research to generate the new
  knowledge required for the practice of evidence-based health care. Centre members are
  involved in individual RCTs; the development of clinically useful measures of cost
  effectiveness; research into the clinical competence and clinical information needs of
  clinicians; research into the precision and accuracy of specific elements of the medical
  history and physical examination; and collaborative research on the usefulness of bench
  research results as diagnostic tests, prognostic markers, and preventive/therapeutic/
  rehabilitative or palliative interventions. The Centre's Director also chairs the Steering
  Group of the Cochrane Collaboration (see Section 3.5 below), and the Centre is closely
  involved in the development of Cochrane Collaboration Methods Working Groups.

The Centre's main vehicle for disseminating sound clinical research is the monthly journal
Evidence-Based Medicine, launched in October 1995 as a UK version of the American *ACP
Journal Club*. The latter is a bi-monthly supplement to the *Annals of Internal Medicine*,

launched in 1991 and accessible online through the Centre for Evidence-Based Medicine at the URL: **http://www.acponline.org/journals/acpjc/jcmenu.htm** It is also available as *ACP Journal Club on Disk* from Customer Service Center, American College of Physicians, Independence Mall West, Sixth Street at Race, Philadelphia, PA 19106 and now includes abstracts, commentaries, correspondence and EBM Notes from *Evidence-Based Medicine*.

Evidence-Based Medicine, which is published as a joint venture by the American College of Physicians and the BMJ Publishing Group, screens over 50 journals each month for articles on diagnosis, prognosis, therapy, aetiology, quality of care and health economics[66]. It focuses on articles which are relevant to medical practice and adhere to rigorous methodological standards, usually identifying some 12 each month. Half the contents come from the internal medicine selections of the *ACP Journal Club*, while the rest consists of the most important clinical studies in the fields of general practice or family medicine, surgery, obstetrics and gynaeocology, paediatrics and psychiatry.

Articles appear in the form of one page structured abstracts introduced by declarative titles which summarise the key elements and clinical conclusions of the original research articles. Each abstract is accompanied by a commentary from an experienced clinician in the field who places the article under review in the context of other related research, identifies any important scientific strengths or weaknesses, and indicates clinical applications. *Evidence-Based Medicine* also contains editorials of general interest on EBM topics, educational material on how to practise EBM, and selected systematic reviews and reports from the Cochrane Collaboration and the NHS Centre for Reviews and Dissemination. David Sackett describes this new journal as publishing 'the gold that intellectually intense processes will mine from the ore of about 100 of the world's top journals'. As such it is a key resource for any doctor or health care manager seeking to practise the principles of EBM.

(66)
ON the need for evidence-based medicine
Sackett, D L; Haynes, R B
Evidence-Based Medicine, Nov/Dec 1995 1(1) pp5-6
Outlines the reasons for the launch of *Evidence-Based Medicine* as a means of bringing 'the best evidence from clinical and health care research to the bedside, to the surgery or clinic, and to the community'. For comment on the criteria for the inclusion of material in the journal
See also: Purpose and procedure (Evidence-Based Medicine, Jan/Feb 1996 1(2) pp34-35)

3.2 Health Technology Assessment Programme

National Coordinating Centre for Health Technology Assessment
Wessex Institute for Health Research and Development
Highcroft
Romsey Road
Winchester
Hampshire SO22 5DH

Tel: 01962 863511
Fax: 01962 844759
URL: **http://www.soton.ac.uk/~wi/hta/ncchta.html**

As noted in Chapter 2, the Health Technology Assessment Programme was set up as a result of the Peckham Report on NHS R&D in 1993, and evaluates the use of devices, equipment, drugs and procedures across medical, nursing and other health care specialities. On 1 June 1996 the management, support and development of the Programme were consolidated in a new National Coordinating Centre for Health Technology Assessment based in the Wessex Region which took over the responsibilities of the Scientific Secretariat (established in Wessex in 1994) and the Administrative Secretariat at the Department of Health. Though its primary location is in Wessex, the NCCHTA is a collaborative organisation involving the Wessex Institute for Health Research and Development at the University of Southampton, and the University of York's Centre for Health Economics and Department of Health Sciences and Clinical Evaluation. The NHS Centre for Reviews and Dissemination, also at York, provides information support on a contractual basis.

The Programme's October 1995 report estimates that over a million interventions using a wide variety of technologies are carried out in the NHS every day, and argues that proper assessment of their value is essential to ensure that clinically and cost effective technologies are promoted and ineffective ones avoided[53, Ch.2]. The role of the Programme is to identify research needs in 'areas of uncertainty in the service' on the basis of an evaluation of evidence, a process which will then have a direct impact on the NHS R&D programme.

Potential topics for study are identified through extensive consultation with those working in and using the health service, and checked against existing sources of research information including the NHS National Research Register and the databases maintained by the NHS Centre for Reviews and Dissemination and the Cochrane Collaboration. This process can lead to several hundred possible topics for research, although there are explicit boundaries to the definition of a health technology:

- 'The health technology under assessment should be clearly defined, sufficiently stable, and be comparable with competing health technologies and/or no intervention, i.e. is there a clear question to answer?
- Evaluating new applications of health technologies is part of health technology assessment but work aiming to develop (new) applications of health technologies would not normally fall under the definition of health technology assessment.
- The end point of the assessment should be improved patient outcomes and/or a more cost-effective means of achieving comparable outcomes.
- Work aimed solely at auditing established good practice or current practice would not normally fall under the definition of health technology assessment.'

Priorities are ranked systematically by the Standing Group on Health Technology and six expert advisory panels on the basis of a set of agreed criteria. For example, 'what are the benefits from an assessment in terms of improved outcomes for patients...improvements in (population-based) cost effectiveness to the NHS, better targeting of service [or] methodological gains'. Priority topics are placed in one of three broad bands:

- Band A: topics of high importance to the NHS in which the potential rate of return would be the greatest. There should be a proactive effort to secure funding for these projects if the normal mechanisms prove unsuccessful. A Commissioning Group for Technology then advises on how best to advance work in these areas.
- Band B: topics of high importance to the NHS, but not meriting proactive effort to secure funding over and above the usual channels.
- Band C: topics of importance to the NHS, but not currently of high enough priority to merit funding. The list of Band C topics is disseminated to other funding bodies to influence responsive research funding decisions.

By September 1995, 54 projects had been funded at a cost of nearly £6 million. Brief details of these are included in the October 1995 report of the Programme.

3.3 NHS Centre for Reviews and Dissemination

University of York
Hesington
York YO1 5DD

Tel: 01904 433634 (General Enquiries); 01904 433707 (Information Service); 01904 433648 (Publications)
Fax: 01904 43361
Email: revdis@york.ac.uk
URL: **http://www.york.ac.uk/inst/crd/welcome.htm**
Telnet address: **nhscrd.york.ac.uk** (user ID and password are both crduser)

The NHS Centre for Reviews and Dissemination was established in January 1994 at the University of York as one of the three elements of the Information Strategy set up to support the NHS R&D strategy. The others are the UK Cochrane Centre (see Section 3.5 below) and the NHS Projects Registers System. The latter is a coordinated network of PC-based research project registers designed to support efficient research management by identifying unwanted research duplication in NHS research; providing decision support information for NHS managers commissioning new research or using existing research; providing a basis for accounting for expenditure on research; and providing input to research reviews and meta-analyses. Access to the Registers, which contain more than 5,000 records of current NHS research, is via the regional NHS directorates although a central coordinating unit oversees development. For more details contact Project Registers Co-ordinating Unit, Vega Group plc, 2 Falcon Way, Shire Park, Welwyn Garden City, Hertfordshire A27 1TW. Tel: 01582 461678.

The Centre for Reviews and Dissemination (CRD) is funded by the NHS Executive and the Health Departments of Scotland, Wales and Northern Ireland, but is independent of government. It cooperates very closely with the UK Cochrane Centre, its 'sibling organisation' with which it has substantial areas of common interest, and with other EBM institutions. It also serves as the UK member of the International Network of Agencies for Health Technology Assessment. Unlike the UK Cochrane Centre which responds to the interest of individuals in undertaking systematic reviews in particular areas, CRD focuses specifically on areas of priority to the NHS. It has a small staff who undertake, or commission and support experts to undertake, reviews and meta-analyses in selected areas, and it focuses heavily on the dissemination of information to enhance decision making[67].

Other functions include liaising with NHS decision makers to prioritise reviews, and the questions raised in reviews; helping to raise the general standard of reviews carried out for the NHS; encouraging research-based practice in the NHS by working with health care professionals, particularly nurses and other therapists in professions allied to medicine who are active in practice and service development; collaborating on research into better methods of disseminating and implementing research evidence, and on research into methods of reviewing the literature; providing an information and enquiry service on reviews and economic evaluations for health care professionals, purchasers and providers, NHS managers, information providers, health service researchers and consumer organisations; and conducting research on the provision of research-based information on the effectiveness of health care to health service users.

CRD maintains two publicly available databases which are a key resource for the EBM movement. Both contain online help, and a more detailed user guide is available from CRD:

- **Database of Abstracts of Reviews of Effectiveness (DARE)** – this includes records of good quality research reviews of the effectiveness of health care interventions, and the management and organisation of health services. These are identified from a variety of sources by trained reviewers and information staff within CRD who evaluate them according to a set of quality criteria and provide detailed structured abstracts. DARE also includes other types of records including source records for reviews which are useful but do not meet all the CRD quality criteria; abstracts of technology assessment reports from health technology assessment agencies from around the world; abstracts of reviews that have appeared in the *ACP Journal Club* and *Evidence-Based Medicine*; records flagging the existence of Cochrane Reviews; and older reviews identified by the UK Cochrane Centre. DARE is available online, on disk and on CD-ROM as part of the Cochrane Library (see Section 3.5 below).
- **NHS Economic Evaluation Database** – this includes abstracts of published economic evaluations of health care interventions, and updates the *Register of cost effectiveness studies* published by the Department of Health in 1994[131, Ch.4]. Records include a structured summary, an assessment of the quality of the studies, and details of any practical implications for the NHS. Cost-benefit analyses, cost-effectiveness analyses and cost-utility analyses are identified by CRD health economists and information staff from a variety of sources, and evaluated according to set criteria. Some records relating to costing studies, methodological articles or reviews are included with basic bibliographical details only.

In addition, CRD is currently developing a register of on-going systematic reviews carried out and/or commissioned both inside and outside the NHS. The aim is to improve the coordination and commissioning of reviews and prevent unnecessary duplication of research; promote collaboration between researchers and organisations working in similar fields; help CRD to identify systematic reviews for possible inclusion in DARE; improve the information service CRD can offer to enquirers; and provide information on systematic review methodology. Information has been collected initially by searching the NHS Research Register and other databases of on-going research, and by contacting NHS bodies and other organisations in the UK known to be involved in commissioning or carrying out systematic reviews. Although this database is not yet publicly accessible, CRD information staff will carry out searches for enquirers.

The dissemination of effectiveness information in a variety of user-friendly formats is a major part of CRD's work, and it produces a range of hard copy publications:

- *CRD Reports* – six of these are available covering the care of critically ill children, the relationship between the volume and quality of health care, the effectiveness of health service interventions to reduce variations in health, ethnicity and health, guidelines for those carrying out or commissioning systematic reviews, and guidelines for those reporting cost-effectiveness studies. The systematic reviews guidance is also available online at the URL: **http://www.york.ac.uk/inst/crd/report4.htm**
- *Effectiveness Matters* – these current awareness bulletins are available free on subscription from CRD and provide summaries of the clinical and cost effectiveness of particular health care interventions for both practitioners and decision makers. Topics covered include aspirin and myocardial infarction, and heliobacter pylori and peptic ulcer.
- *Effective Health Care Bulletins* – this bi-monthy bulletin is produced jointly with the Nuffield Institute of Health at the University of Leeds and published by Churchill Livingstone (Contact Churchill Livingstone Subscriptions Department, PO Box 77, Fourth Avenue, Harlow, Essex CM19 5BQ). Each *Bulletin* consists of a systematic review and synthesis of research on the clinical effectiveness, cost-effectiveness and acceptability of health service interventions carried out by a research team using established methodological guidelines and advice from expert consultants. Topics covered to date are: Vol 1 No 1: screening for osteoporosis to prevent fractures; Vol 1 No 2: stroke rehabilitation; Vol 1 No 3: the management of subfertility; Vol 1 No 4: the treatment of persistent glue ear in children; Vol 1 No 5: the treatment of depression in primary care; Vol 1 No 6: cholesterol: screening and treatment; Vol 1 No 7: brief interventions and alcohol use; Vol 1 No 8: implementing clinical practice guidelines; Vol 1 No 9: management of menorrhagia; Vol 2 No 1: the prevention and treatment of pressure sores; Vol 2 No 2: benign prostatic hyperplasia: treatment for lower urinary tract symptoms in older men; Vol 2 No 3: management of cataract; Vol 2 No 4: preventing falls and subsequent injury in older people; Vol 2 No 5: preventing unintentional injuries in children and young adolescents; Vol 2 No 6: management of primary breast cancer. Future *Bulletin* topics will include coronary heart disease, stroke prevention in the elderly, total hip replacement, alcohol and substance abuse in younger people, mental health promotion, back pain, prevention of infections in surgery, sleep apnoea and obesity.

(67)
EVIDENCE-BASED practice: the role of the NHS Centre for Reviews and Dissemination
Glanville, J
Health Libraries Review, Dec 1994 11(4) pp243-51, 17 references
The CRD Information Services Manager discusses the background to the establishment of the CRD, describes its systematic review work, and discusses information dissemination and information services based on the Centre's databases. Includes comment on the important role played by librarians and other information professionals in the promotion of evidence-based practice. For more on CRD
See also: Research intelligence for policy and practice: the role of the National Health Service Centre for Reviews and Dissemination, by T A Sheldon (Evidence-Based Medicine, Sep/Oct 1996 1(6) pp167-68)

3.4 UK Clearing House for Information on the Assessment of Health Outcomes

Nuffield Institute for Health
71-75 Clarendon Road
Leeds LS2 9PL

Tel: 0113 233 6974 (HELMIS Manager, Information Resource Centre: for priced searches of the databases)
URL: **http://www.leeds.ac.uk/nuffield/infoservices/UKCH/home.html**

Health outcomes are the effects on health of any type of process including NHS and private health care but also housing, social services and employment[68]. Information on health outcomes is needed to eliminate poor and/or unnecessary practice, and to promote good practice based on the best evidence. In addition it can help to improve accountability within the health service and empower consumers. Such information is highly valued by practitioners and managers, and complements that provided by the systematic review of RCTs and other scientific research.

The UK Clearing House was set up jointly in 1992 by the Department of Health and the health departments in Northern Ireland, Wales and Scotland to act as a central reference point for work on outcome measurement, drawing on in-house experience and work being carried out in the NHS by practitioners in the field[69]. Recent funding decisions led to its closure at the end of March 1997 but outcomes measurement is still central to the R&D activities of the Nuffield Institute for Health, and resources have aleady been secured for selective topic-based review work. It will also continue to offer help with organising local and regional workshops on issues in outcomes measurement. Moreover, the databases produced by the Clearing House remain available via the Internet, while its publications can be obtained from UKCHHO Publications at the Nuffield Institute for Health.

- *Outcomes Briefing* – each *Briefing* is devoted to a particular theme, with additional workshop and conference reports, invited contributions and details of current literature. The eight titles are: An introduction to measuring health outcomes; Multidimensional profiles; Searching the outcomes literature; Review of the SF-36; Exploring outcomes within routine practice; Broadening the base for outcomes measurement: criteria for selecting measures; Outcomes within clinical audit; and Outcomes, patients and carers.
- *Outcomes Measurement Reviews* – five of these are available dealing with outcome measurement on a topic basis. Titles are Measuring the outcomes of total hip replacement through the commissioning process; Assessing and monitoring the health outcomes of alcohol misuse; Measuring the health outcomes of adult asthmatics; Measuring the health outcomes of diabetes care; and Measuring the outcomes of stroke care.
- *Outcomes Measurement Bibliographies* – four titles are available in this series of annotated bibliographies entitled An introduction to measuring health outcomes; Measuring the health care outcomes of adult asthmatics; Measuring the health care outcomes of diabetes care; and Measuring the outcomes of stroke care.
- **Outcomes Activities Database** – this is available via the World Wide Web or on floppy disk (DOS and Windows versions), and its primary aim is the development of a network of those involved in health outcomes measurement. Thus the quality of projects is not assessed before inclusion in the database although they are classified into

one of five categories: the application of outcome measures within contracting; outcomes criteria and measures used in clinical audit; evaluations of clinical interventions within routine practice; projects developing or testing outcome measures; and projects exploring the effectiveness and outcomes of health care interventions. Links are also provided to similar databases in other European countries. The database will continue to be maintained as long as the Nuffield Institute for Health is informed of new projects and, with this in mind, a form will be established on the website and published in relevant journals.

- **Outcomes Database of Structured Abstracts** – provides access to critiques of papers discussing the development and testing of instruments in the areas of asthma, diabetes and stroke.
- **Outcomes Literature Database** – provides access to selective literature dealing with outcomes measurement.

The website has been updated and extended to provide a comprehensive list of Clearing House publications and details of how to order them; summaries of publications; access to the databases; links to other websites dealing with outcomes, effectiveness and quality management; and information sheets on a variety of topics. These cover defining an outcome; clarifying desired outcomes; exploring the outcomes of diabetes care, fractured neck of femur, severe mental illness and stroke management; reviewing outcome measures; and searching for information on outcomes measures and measurement on MEDLINE and CINAHL.

(68)
MEASURING health and medical outcomes
Jenkinson, C (editor)
UCL Press: London, 1994. 215pp (Social Research Today 3)
For more comment on outcome measurement in the NHS, including comment on the role of outcomes assessment in improving clinical effectiveness from the head of the NHS Central Health Outcomes Unit which was set up in 1993 to encourage and coordinate the development and use of information on health outcomes
See also: Outcome measurement in the NHS, by L Macdonald and others (University of Aberdeen, Department of Public Health, Mar 1995. 138pp)
The role of outcomes assessment in improving clinical effectiveness, by A Lakhani (In: Clinical cost effectiveness: from guidelines to cost-effectiveness, edited by M Deighan and S Hitch. Earlybrave Publications Limited, PO Box 3165, Brentwood, Essex CM13 1TL, Oct 1995 pp39-48)

(69)
THE DEMAND for and use of outcomes information
Warburton, A and others
Health Libraries Review, Dec 1994 11(4) pp253-61, 11 references
A useful summary of the background to, and work of, the UK Clearing House on Health Outcomes which also includes the results of a survey of enquiries received. For comment on the establishment of the Clearing House, and for its valedictory publication bringing together the papers presented at the Third International Meeting of the European Clearing Houses on Health Outcomes (ECHHO)
See also: Establishment of UK clearing house for assessing health services outcomes, by A Long and others (Quality in Health Care, Jun 1992 1(2) pp131-33)

Health outcomes and evaluation: context, concepts and successful applications, edited by A Long and E Bitzer (Nuffield Institute for Health, UKCHHO Publications, 71-75 Clarendon Road, Leeds LS2 9PL, May 1997)

3.5 UK Cochrane Centre/Cochrane Collaboration

Summertown Pavilion
Middle Way
Oxford OX2 7LG

Tel: 01865 516300
Fax: 01865 615311
Email: general@cochrane.co.uk
URL: **http://hiru.mcmaster.ca/cochrane/centres/uk/default.htm** (the UK Cochrane Centre does not have its own web page, although it does have an FTP server at ftp://ftp.cochrane.co.uk)
URL: **http://hiru.mcmaster.ca/cochrane/default.htm** (this is the Cochrane Collaboration home page)

The UK Cochrane Centre, named after Archie Cochrane, was established in October 1992 as one of the three components of the Information Strategy set up to support the NHS R&D strategy (the others being the NHS Centre for Reviews and Dissemination and the NHS Projects Register)[70]. It is funded by the NHS R&D Programme for England; the Scottish Home and Health Department; NHS Wales Office of Research and Development; the Department of Health and Social Services, Northern Ireland Office; the Anglia and Oxford Region; the European Commission's BIOMED 1 programme; the Nuffield Provincial Hospitals' Trust; and the Medical Insurance Agency. Since its establishment an international Cochrane Collaboration, of which the UK Cochrane Centre is an integral part, has evolved including members from Australasia, Canada, the USA, the Netherlands, Italy and the Nordic countries.

During its second period of funding (1994-1999) the UK Cochrane Centre's objectives are:

- To identify module editors and reviewers in the UK and elsewhere who are able and willing to prepare and maintain systematic reviews of the effects of health care, with particular reference to NHS priorities for research-based information.
- To support module editors, reviewers and others in the UK who are contributing, or wish to contribute, to the Cochrane Collaboration.
- To develop and maintain a management system for use throughout the Cochrane Collaboration for incorporating edited review group modules within the Cochrane Database of Systematic Reviews.
- To collaborate with the NHS Centre for Reviews and Dissemination, and other groups, to make information contained in systematic reviews available through the Database.
- To explore, with others, ways in which people using health services could help to plan and develop protocols for Cochrane reviews.
- To contribute to the infrastructure, coordination and development of the Cochrane Collaboration in association with other Cochrane Centres and the Steering Group.

Systematic reviews are prepared by groups of self-nominated individuals (typically including health professionals, methodologists and consumers) who share a common interest in a health care effectiveness issue. The UK Cochrane Centre helps to identify other potential group members from around the world, and provides support for group facilitators to plan and run an exploratory meeting for all those who have expressed an interest. As a result of this and subsequent meetings, teams of group members voluntarily take on, for the rest of their careers, the task of preparing and updating systematic reviews of the effects of specific health care interventions for specific conditions. The voluntary basis of the Collaboration is a particularly striking feature, but whether it will prove robust enough to survive beyond the early years of enthusiasm has yet to be tested.

The teams are formally registered by the Cochrane Collaboration as Collaborative Review Groups once they have worked out agreed procedures and protocols for undertaking the work which is carried out with the advice and support of the UK Cochrane Centre. Team members will, for example, be trained in both the searching of computerised databases and in the manual searching of journals to ensure that they track down as many RCTs relevant to their area of study as possible. During this process, they will come into contact with two further types of Cochrane group:

- Fields – these groups, which are currently evolving, share common interests in groupings of health care consumers (e.g. older people); a branch of health care (e.g. public health); or a broad category of health care intervention (e.g. physiotherapy).
- Methods Working Groups – these groups work to organise and disseminate the work of methodologists who have come together to improve the quality of systematic reviews, and cover areas such as manual searching of journals, and the use of specialised software in preparing, analysing, maintaining and presenting systematic reviews.

The Cochrane Collaboration was launched at the first Cochrane Colloquium at Oxford in October 1993. This has been followed by three further colloquia in Hamilton, Ontario, Oslo and Adelaide whose proceedings are available via the Collaboration's home page. They have included a wide range of practical workshops and papers on the many processes involved in systematic reviews and meta-analyses, as well as presenting the results of specific reviews in many areas. From April 1995, the Cochrane Collaboration has been incorporated as a public limited company with charitable status, under the lead of a Steering Group chaired by David Sackett, and draws on many sources of funding from governments, research agencies, universities, charities and individuals.

The output of the Cochrane Collaboration – the Cochrane Library – consists of four databases:

- **Cochrane Database of Systematic Reviews (CDSR)** – this includes nearly 100 systematic reviews in the areas of pregnancy and childbirth, subfertility, stroke, schizophrenia and parasitic diseases. In addition it includes 94 protocols in the areas of acute respiratory infections, airways, diabetes, musculoskeletal injuries, neonatal care, and peripheral vascular diseases. Smaller, specialised databases are derived from the CDSR and published on disk.
- **Database of Abstracts of Reviews of Effectiveness (DARE)** (see Section 3.3 above).
- **Cochrane Controlled Trials Register (CCTR)** – includes bibliographical details of over 100,000 controlled trials identified by Cochrane Collaboration contributers including many not currently listed in MEDLINE, EMBASE and other standard bibliographic databases.

- **Cochrane Review Methodology Database (CRMD)** – a bibliography of articles on the science of research synthesis and practice aspects of preparing systematic reviews.

The databases are available in CD-ROM for Windows format (containing CDSR, DARE, CRMD and complete CCTR); or 3[AB] inch disk for Windows (containing CDSR, DARE, CRMD and 1995-96 CCTR). The Anglia and Oxford Region of the NHS has produced a training guide for the Library which can be downloaded and includes an introduction to the definition and value of systematic reviews; the work of the Cochrane Collaboration and the NHS Centre for Reviews and Dissemination; a pictorial guide to the databases together with simple exercises to encourage exploration; and an explanation of odds-ratios and their use. The guide is designed for individual or group use.

Detailed information about the Collaboration is contained in its key working document, the regularly updated *Cochrane Collaboration handbook* which is produced in both electronic and paper formats (available from **http://hiru.mcmaster.ca/cochrane/handbook/ default.htm**). Each of its six sections is the responsibility of a separate editorial team under the overall control of Andrew Oxman and David Sackett, and cover the Cochrane Collaboration; establishing and supporting collaborative review groups; representing the interest of fields; the Cochrane Centres; establishing and maintaining registers of RCTs; and preparing and maintaining systematic reviews. Sections also include detailed bibliographies on many EBM issues.

(70)
PREPARING and updating systematic reviews of randomized controlled trials of health care
Chalmers, I and others
Milbank Quarterly, 1993 71(3) pp411-37
Looks at this issue in the context of the UK Cochrane Centre's work on pregnancy and childbirth. For more on the Cochrane Collaboration
See also: The Cochrane Collaboration deserves the support of doctors and governments, by F Godlee (British Medical Journal, 15 Oct 1994 309(6960) pp969-70)
The Cochrane Collaboration: preparing, maintaining and disseminating systematic reviews of the effects of health care, by L Bero and D Rennie (Journal of the American Medical Association, 27 Dec 1995 274(24) pp1935-38)

4. GETTING EVIDENCE INTO PRACTICE

Provision of good quality information on clinical and cost effectiveness is only the first stage in creating an evidence-based health service. In the NHS, the work of the Cochrane Centre, the NHS Centre for Reviews and Dissemination, the Health Technology Assessment Programme and the many other bodies involved in meta-analysis and review, is designed to give access to externally generated sources of information on effectiveness. Clinical audit, outcome assessment and other performance monitoring initiatives perform the same function within the organisation. However, none of this is of any practical value unless new information is translated into changes in practice where these are found to be necessary[71].

Thus mechanisms have been set up to try and ensure that effectiveness information from both internal and external sources is taken seriously by health care purchasers and providers. Health authorities and trusts are now formally accountable to the NHS Executive regional offices for their performance in improving clinical effectiveness through mechanisms such as clinical guidelines, continuing professional development, and targeted post-graduate education. However, important though such managerial initiatives are, their success is ultimately dependent on the willingness of individual health care professionals (who take the vast majority of decisions made about health care in the NHS) to change their attitudes and behaviour.

While EBM may not necessarily be a paradigm shift in medical practice, it undoubtedly involves disruption both to traditional ways of working and to traditional authority structures. Crucially, it is also being promoted at a time of unprecedented change in the health service as a result of the introduction of the internal market reforms, more rigorous performance monitoring and an increasing emphasis on patient empowerment[72]. All of these developments have served to increase managerial control over doctors and other health care professionals, and challenge their traditional autonomy. If EBM, as an integral part of the health service revolution, is to succeed, those whose lives are disrupted need to be convinced that the benefits outweigh the costs.

Inertia and hostility are almost universally experienced by organisations which seek to promote radical cultural change, hence the frequently expressed disappointment of management gurus and senior executives that business process re-engineering, Total Quality Management or other management innovations fail to deliver the expected benefits. Change imposed from above or from outside is often perceived – quite understandably – as a threat, and is resented for its implicit (sometimes explicit) criticism of current ways of doing things. If those current ways of doing things are supported by strong professional values and hallowed by high social status, resentment and resistance may be all the more likely. The mechanisms of a new order may be put in place, and people may go through the motions of change, but real commitment to innovation may be in short supply. For this reason, it is appropriate to look first at some of the views expressed by doctors about the EBM revolution before moving on to some of the ways in which it is being implemented.

4.1 The professional debate

Strong professional values, high social status and – on occasion – resistance to external interference are notable characteristics of the medical profession and ones which may sometimes conflict with the EBM agenda. The evidence-based model of medical practice may well be unattractive to those doctors whose self-esteem and professional status are based on pride in their personal skills and intuitive understanding, painfully acquired through years of training and experience. To be told that these are no longer enough may well provoke resentment and hostility, especially among senior doctors who might also feel that their authority over juniors is being challenged. If the keystone of professional competence is no longer to be experience but the ability to appraise and use external evidence, much of the rationale for rigid professional hierarchies based on age and seniority disappears.

The temptation to reject EBM as 'cook-book' medicine can be strong among those who feel that their professional competence is being questioned and their clinical freedom threatened by people who are frequently no longer in regular contact with patients. However much the EBM community may stress the fact that traditional skills and practices remain a key part of professional competence, there will be those who feel that the venerable art of medicine is being unfairly devalued by people who are no longer practising it on a day-to-day basis[73]. 'Diagnosis by scientific literature', they feel, is the antithesis of what medicine should be about[74]. It ignores the uncertainty inherent in much of medical practice, and arbitrarily excludes the knowledge and understanding that can be provided by what have been called the 'non-biological arts'[75]. Not all that is measurable is of value, and not all that is of value can be measured, according to this view.

An editorial in *The Lancet* in September 1995 expresses what may be the view of a significant number of doctors when it complains that 'the voice of evidence-based medicine has grown over the past 25 years or so from a subversive whisper to a strident insistence that it is improper to practise medicine of any other kind'[76]. As a result it threatens to alienate doctors 'who would otherwise have taken many of its principles to heart'. Andreas Polychronis and his colleagues, writing later in the *Journal of Evaluation in Clinical Practice,* are equally blunt, accusing the protagonists of EBM of being 'arrogant, seductive and controversial' in their attempts to promote as the new medical orthodoxy what is little more than a useful addition to the toolbox of clinical practice[77].

David Sackett's angry response to *The Lancet* argued that 'it is easy to have enthusiasm about evidence-based medicine mistaken for elitism and arrogance', especially by a profession 'accustomed to warding off attacks from the outside and even less ready to respond to clamour for change from within its own ranks'. No doubt this has been interpreted as further evidence of EBM pushiness by his critics including one writing in response to the same editorial to complain that 'the steps and recommendations of the evidence-based medicine acolytes reek of obfuscation and platitudes'. Another condemns David Sackett and his colleagues for 'their arrogance, their jargon, and their penchant for denigrating others'. However, yet another identifies what may be the real nub of the matter for conservative members of the profession, jealous of their status: 'knowledge based on a scientific discourse is democratic and open to debate, and knowledge based on expertise is oligarchic and closed. It is not surprising that doctors argue against evidence-based medicine with such vehemence'.

Ecological fallacies

It would be unfair, however, to suggest that opposition to EBM within the medical profession is solely based on a desire to preserve traditional authority structures. The most vociferous critics challenge one of the most fundamental tenets of EBM – the use of meta-analyses of RCTs or other reputable scientific evidence as the basis for clinical practice – arguing that the epidemiological emphasis of EBM is mistaken[78]. The results of RCTs, and particularly meta-analyses of several RCTs, provide information only at the population level and to apply them to the care of individual patients is to fall into the classic ecological fallacy trap. Group averages tell the practitioner nothing about the causal processes in the indivduals who compose the group, and may mask research errors and inconsistencies[79]. The uncritical application of this kind of information in everyday practice can be both misleading and dangerous, while the narrow selection criteria employed in the recruitment of patients to trials mean that the results of RCTs are often far less applicable to the general patient population than those of less scientifically rigorous studies[80].

Brian Haynes, director of the Canadian Cochrane Centre (and, therefore, an important member of the EBM community) uses the example of thrombolytic therapy during acute myocardial infarction to illustrate some of the problems of using RCTs[81]. Although many trials have demonstrated the effectiveness of this therapy in saving lives, the narrow criteria used to select patients for inclusion means that as many as 90% of those presenting with heart attack from the general population may be ineligible for this particular form of treatment. Identifying the 10% who may benefit takes time and this is rarely available to the clinician on the ward or in the accident and emergency department. Victims of heart attack often delay seeking treatment, and thrombolytic therapy, if it is to be effective, must be applied within hours. Finally, despite all the trials, there still remains considerable debate about the effect on responsiveness to thrombolytic therapy of a wide range of other factors such as age, infarct location and extent, and the presence of other medical conditions.

Humane and effective medicine, according to one commentator, 'is underpinned by two ethical values. Firstly, there is the value that we attach to the belief that other people matter and that the meaning of any disease is personal and unique for them; a physician's skill is to understand and act on this. Secondly, there is the value that says you cannot possibly advise, or even respect, others unless you are in possession of the facts. And the ethic here is to know and be able to examine dispassionately the results of any intervention'[82]. Thus the results of clinical trials, and other scientific evidence, can do no more than inform practice in those areas where such information is available. Any attempt at 'dictating medical activity according to algorithms and systematic reviews does violence to the way medicine is practised' by denying the subtlety and sophistication of its non-scientific skills. Such views, according to a recent American analysis, may be widely held and provide a powerful intellectual justification for those who seek to preserve the traditional status of clinical reasoning based on experience and respect for the individual patient[83].

Even among those many doctors who accept the value of meta-analyses in theory, there may be a lingering suspicion that they cannot really be trusted to inform the care of individual patients[84]. The Coopers and Lybrand survey of GPs reported in *Consider the evidence* shows that 95% would be persuaded to change their practice on RCT evidence[72]. However, 'real world' data from clinical outcome studies is slightly more important (96%), while the doctor's own observations and those of patients are also cited as important generators of change. Voices are also being raised in support of qualitative research in the health care field, and a literature review on this subject is due to be published by the Health Technology Assessment programme later in 1997[85].

The EBM community responds to the ecological fallacy argument with a vigorous defence of meta-analyses and a denial that EBM is 'merely the mindless application of the results of megatrials'. The results of a single RCT (or other study) can be valuable, but the systematic analysis and synthesis of the results of several well conducted and replicable studies are likely to give a better overall view of the effectiveness of a given medical intervention. Equally, though RCTs occupy an important place in EBM, they are not the only sources of information which can be used, and nor are these other sources (including cohort, case control and cross sectional studies) 'heavily discounted' as the critics suggest. The EBM community garners evidence from a wide range of appropriate sources, and will expand these as time and resources permit. It also emphasises the crucial importance of systematic evaluation of observational data by the individual clinician in the light of the characteristics and preferences of real patients. EBM is not mindless cook-book medicine, but a means of enhancing traditional clinical skills through better information.

Shortages and duplication of evidence

Regardless of the fact that RCTs are not the only source of evidence used in EBM, they are clearly very important, and a further obstacle to the development of an evidence-based health service which the critics identify is a shortage of this gold standard information. Although RCTs of new drug therapies are now routine, there are many areas of diagnosis and treatment which still lack the rigorously evaluated and up-to-date primary evidence on which an EBM approach might be based. Indeed, the EBM revolution itself is serving to highlight this sobering fact. Coverage of the non-medical, but equally important, interventions practised by nurses[86], physiotherapists[87] and many other health care professionals is extremely limited although attempts are being made to improve matters. Gaps in the medical field include much of general practice[88] and non-acute care, while the coverage of acute care remains patchy, with several important specialties such as haemotology, otolaryngology and neurology very poorly served.

An article in *The Lancet* in April 1995 refers to the 'many grey zones of medical practice where the evidence about risk-benefit ratios of competing clinical options is incomplete or contradictory'[89]. Despite the public perception of medicine as a high-tech, knowledge-intensive business, 'clinical medicine seems to consist of a few things we know, a few things we think we know (but probably don't), and lots of things we don't know at all'. Nonetheless, areas of overlap in sources of information are already beginning to appear so that, for example, a health care purchaser or provider wanting information on the management of coronary heart disease currently has a choice of at least four different sources. Information shortages exist, paradoxically, side by side with information overload.

Rigorous scientific evidence may always be in relatively short supply in medicine despite the scores of trials now published every month. There are many areas in which double-blind RCTs are practically impossible (surgery, for example) or ethically undesirable, and many clinical situations in which the doctor has little to call upon beyond personal experience and a knowledge of pathobiology[90,91]. However, this does not mean that the principles of EBM cannot be applied in the grey zones where uncertainty reigns. Indeed all zones may be grey when the doctor is faced with an individual patient, even where scientifically validated evidence from RCTs or other reputable sources is available. It is primarily the *systematic* nature of EBM which distinguishes it from the traditional model of medical practice, not its emphasis on scientifically validated evidence, and its skills are as applicable to the grey zone as to any other.

Shortages of time and skills

For those doctors who do wish to practise EBM, there are major difficulties as a result of the sheer weight of potential evidence that needs to be reviewed – and this despite the fact that many areas of practice lack the kind of rigorous, gold standard, evidence on which such an approach might be based. Information overload may apply even to the majority who rely on the evaluation skills of others producing systematic reviews. David Sackett and his colleagues are convinced that the practice of EBM by individual doctors or clinical teams is possible, but Chapter 3 shows that the number of organisations active in the EBM field is growing rapidly and they produce a bewildering variety of newsletters, reports, guidelines, web sites and databases which document meta-analyses and reviews. No one source is sufficient to cover every need, and simply keeping track of those that might be relevant to current awareness and problem-solving can be a major task in itself.

It also requires access to computer hardware and software, and – if the practitioner needs to carry out his or her own appraisals of the literature – the ability and time to search online bibliographic databases, CD-ROMs and/or the Internet. Basic computer literacy may now be universal among medical students and junior doctors, but sophisticated skills can be needed to overcome the deficiencies of MEDLINE and other databases which lack both comprehensive coverage and fool-proof indexing[92]. Older doctors may find the skills difficult and time-consuming to acquire, although the EBM community argues that the technology, and its use, are not nearly as daunting as many fear. Using electronic sources of information also costs money, not only for the necessary equipment but also to buy CD-ROMs or subscriptions to online services (though access to MEDLINE is provided free by the British Medical Association to members with modems).

Ensuring that knowledge is available at the bedside or in the consulting room when it is needed, and in a form that can be used to the benefit of the patient, is yet another problem. It has been estimated that formulating a question, finding the evidence, appraising it and acting on it take about two hours – clearly not much use in an emergency situation. Complex or unusual clinical problems may require longer searches, and a significant proportion of attempts may end in failure because of the inadequate coverage or indexing of databases, or simply because there is no evidence to find. As a result, many doctors – and some members of the EBM community – doubt whether EBM will ever be of much practical use to front-line medical personnel. It is always likely to be of more value to those doctors involved in medical management, a view which seems to be born out by the Coopers and Lybrand study which includes survey evidence on the EBM attitudes and practices of both GPs and members of the British Association of Medical Managers[72].

Front-line scepticism

In general practice, for example, the following arguments are frequently put forward by doctors who may well feel there are hardly enough hours in the day to give patients the time they need, let alone cope with the paperwork and 'do the educational stuff'[93]:

- **There is no scientific evidence for more than a tiny fraction of decisions made in primary care.**
  This was a legitimate argument ten years ago, according to the EBM community, but less so today. The number of RCTs indexed in MEDLINE under primary care has increased five-fold since 1986 although the total admittedly still remains small and a significant proportion of drug trials involving GPs may fail to yield scientifically valid and clinically relevant findings[94]. Nonetheless, 'a lot of high-quality, relevant evidence is already there, but it remains invisible to most GPs, even those who keep up to date with the mainstream journals'[95].

- **Even if the evidence is there, GPs do not have the time to track it down.**
  They do not need to, say the EBM enthusiasts, because much of the work has already been done by the systematic reviewers and can be accessed through sources such as *Evidence-Based Medicine*, the *ACP Journal Club*, and the *Cochrane Database of Systematic Reviews*. The American *Journal of Family Practice* also has its own journal club accessible via **http://www.phypc.med.wayne.edu/jfp/jclub.htm** which identifies seven to ten key articles each month from some 80 clinical journals in the primary care field, critically appraises them, and makes specific recommendations for clinical practice (known as POEMs – Patient Oriented Evidence That Matters – for short). The Medi-Search Article Archive provided by DocNet, an Internet service administered by UK GPs and available via **http://www.docnet.org.uk/medis/search.html** is a further source which indexes articles from a range of online medical journals. These kinds of initiative have, in the view of the enthusiasts, radically reduced the information overload problem by filtering out poor quality research and focusing on that which is of real value. Even in those many areas where meta-analyses do not exist, the GP can make 'huge strides' simply by setting aside an hour a week for searching and appraising the literature.

- **GPs lack the skills and experience to appraise evidence and determine its applicability in specific clinical settings**
  Many do require training in computer use, literature searching and critical appraisal but acquiring these skills is not nearly as difficult as it sounds according to the EBM community. Indeed, increasing numbers of younger GPs already have them through their undergraduate and postgraduate education. The examinations of the Royal College of General Practitioners have included a section requiring the critical appraisal of a clinical article for some years, while a recent guide to *Evidence based general practice* gives practical advice[96].

- **Relevant evidence cannot be recalled during the consultation when it is needed**
  Overcoming this problem requires a computer on the desk of the GP, the skills to use it, and the availability of reference-managing software that can allow the practitioner to keep the evidence needed to help with the most frequently occurring problems within easy reach[97]. Studies from the USA and elsewhere suggest that growing computer literacy and the development of computer-based decision support systems can have significant effects on clinical performance and patient outcome[98]. Among recent initiatives in this area is the development of CATmaker by the Centre for Evidence-Based Medicine. This prototype Internet tool (also available in a stand-alone application) partially automates the generation of Critically Appraised Topics (CATs) and allows the information to be copied and retrieved at will[99]. The

Department of Health has also invested some £1.5 million in the PRODIGY initiative to encourage rational prescribing in primary care which has been tested over a six month period in some 120 practices. It offers the GP access to three choices of drug treatment for each diagnosis and is claimed to cover 90-95% of all GP consultations. Interestingly, it has come under fire from both the Association of the British Pharmaceutical Industry and patients' groups for being too inflexible and insufficiently responsive to patient demands for more choice on drug treatment[100].

Despite the reassuring tone adopted by the EBM community, there is clearly some way to go before all front line health care professionals accept the evidence-based approach for the rational, effective, easy-to-use process its advocates believe it to be. Even the majority who support it in principle may find the difficulties of putting it into practice insuperable. In the case of GPs, the Audit Commission's 1996 study of *What the doctor ordered* shows that 'most fundholders are not making full use of the increasing body of knowledge about clinical effectiveness to inform their commissioning decisions'[101]. Although the Coopers and Lybrand survey shows that most GPs are quite positive about EBM, they still complain at being 'inundated with paper' and overloaded with information[72].

The devil in the detail

The National Association of Health Authorities and Trusts (now the NHS Confederation) in its 1995 report on *Acting on the evidence*, gives a useful summary of many of the practical problems the interested manager and practitioner (general or specialist) might face in trying to practise EBM, citing issues such as the shortages of high quality evidence in many areas; the fragmentation and overlap between sources of information; variations in the way information is presented; difficulties in getting access to sources (not all are online and available 'at the touch of a finger tip', while Internet use is increasingly subject to frustrating delays); and continuing worries about the quality and currency of clinical guidelines and some meta-analyses[102]. Though the principles of EBM seem robust and sensible to many, the devil lies in the detail of application.

NAHAT's most telling comment focuses on the fact that the evidence-based health care movement appears to be developing 'without much detailed thought or preplanning as to how or by whom that information will be used...at present it is unclear who is expected to use their information, or what they are expected to do with it, or how the information should be presented to maximise its application'. In this context, it is unsurprising that some health care professionals remain to be convinced about its value, while others find that trying to put an evidence-based approach into practice raises all sorts of unforseen problems. For example, a given evidence-based innovation may be accepted as clinically effective, but opposed in practice because it requires retraining, changes in working practices or other costly non-clinical developments. NAHAT cites the deceptively simple example of changing patients' waking times, an innovation which may be of considerable benefit to patient well-being but can have disruptive knock-on effects on virtually all hospital staff.

At the same time, health care providers may be reluctant to implement evidence-based changes until they have proved their value for themselves. They may doubt the standard of proof provided by RCTs or other sources of evidence, however reputable in scientific terms, when it is to be applied to the care of individual patients, and need to verify new interventions for themselves. This is not always possible, and is frequently expensive. The degree to which purchasers are in a position to induce providers to change their practice in the face of new evidence is also in doubt, raising difficult questions as to how far managers can or should dictate to clinicians. Certainly, attempts to impose an evidence-based culture through the contracting system are likely to founder on the resistance of sceptics, and the process needs to be one of widespread consultation and agreement reflected in, rather than driven by, the contracting system[103].

The complexity of the implementation issues which face the NHS in seeking to move towards an evidence-based health care system has been exhaustively examined by one of the expert advisory groups reporting to the Central Research and Development Committee[104]. There is no shortage of potential mechanisms to increase the uptake of research findings by practitioners, but the effectiveness of the implementation process is influenced by a multitude of organisational, cultural and other factors including:

- **'Characteristics of message (research findings) including**

  Scientific quality
  Source
  Content (complexity of information)
  Presentation (e.g. relative vs absolute risk reduction; numbers needed to treat; individual studies vs systematic reviews)

- **Characteristics of players (those influencing practice or being influenced) including combinations of**

  Individual health professionals and managers
  Purchasing organisations
  Provider organisations (trusts and health authorities, including non-executive directors)
  Professional organisations (including royal colleges)
  Industry, e.g. pharmaceutical
  Education providers
  Research information providers
  Researchers
  Public
  Patients
  User groups
  Media
  "Alliance" partners, e.g. local authorities
  Policy makers

- **Characteristics of interventions including combinations of**

  Use of clinical guidelines
  Audit and feedback
  Conferences
  Local consensus processes
  Educational approaches
  Marketing
  Opinion leaders
  Reminders/computerised decision-support
  Patient-mediated interventions

- **Levers, facilitators and barriers including combinations of**

  Availability of resources
  Availability of time
  Financial/contractual
  Statute
  Professional incentives/disincentives
  Cultural/social/organisational norms'

There is, according to the advisory group, 'a paucity of research in this field', although some complementary work is underway. For example, the NHS Centre for Reviews and Dissemination carries out some research into methods of dissemination as part of its work, and a Cochrane Collaboration on Effective Professional Practice is gathering together existing research (mainly in the form of controlled studies) on changing the behaviour of health care professionals and the public[105]. A further report to the CRDC on the primary care interface also considers research-into-practice issues[106], while the Economic and Social Research Council's large scale research programme on innovation may well provide useful insights from areas outside health.

However, considerable research still needs to be done and the group identifies twenty priority areas including the influence of the source and presentation of evidence on uptake; the management of uncertainty and communication of risk by clinicians; why some clinicians, but not others, change their practice in response to research findings; the effectiveness and cost-effectiveness of audit and feedback as a way of promoting the implementation of research findings; the impact of practice guidelines in disciplines other than medicine; and the role of the media in promoting research uptake. Educational issues, discussed in the next section, are also flagged given the importance of new skills and attitudes in promoting change.

Unfortunately – as a much-quoted analysis by Andrew Oxman shows – there are no 'magic bullets' for ensuring effective implementation of research findings[107]. Simply disseminating evidence has been proven to be largely ineffective, but so too have many of the more active strategies. Better dissemination is clearly required according to the Coopers and Lybrand survey which shows 'easy access to EBM summaries' as the second most important lever to influence the uptake of EBM for both GPs and medical managers (following 'clear evidence it would improve patient outcome')[72]. However, there are others of significant, though lesser, importance including direct financial incentives (for GPs at least); discounts in the medical insurance premium; formal recognition of those who routinely practise EBM; and its requirement to gain accreditation to practise or become a 'preferred provider'. Clearly it is essential to take into account not just the quality of the evidence and the efficiency of dissemination, but the complex cultural, organisational and economic environment in which the practitioner operates[108].

It is perhaps not surprising that the latest assessment of clinical effectiveness implementation by the research team that produced *Acting on the evidence* makes it clear that much still remains to be done despite some progress in introducing EBM principles into the NHS[102]. Though clinical effectiveness is high on the policy agenda, the mechanisms by which the Department of Health and the NHS Executive measure and assess performance still tend to emphasise efficiency and economy rather than whether the care provided actually works. Performance requirements such as the Patient's Charter

standards for surgical waiting lists may encourage surgeons to perform less effective
procedures on those who have waited for a long time, rather than focus on more effective
treatments for other patients. Equally, the contracting system values all admissions and
clinic visits equally, regardless of the appropriateness of care provided. Until performance
incentives within the NHS work with, rather than against, the grain of EBM, the claim
that clinical effectiveness principles drive the NHS may be rather a hollow one.

(71)
GETTING research into practice
Roland, M
Journal of Epidemiology and Community Health, 1995 49(3) pp225-26, 32 references
A brief look at the research to practice problem, with comment on UK initiatives at local
and national level.

(72)
**CONSIDER the evidence: the NHS on the move towards evidence-based
medicine**
Felton, T; Lister, G
Coopers and Lybrand, 1 Embankment Place, London WC2N 6NN, Apr 1996. 51pp
Focuses in particular on the EBM attitudes and practices of GPs and medical managers,
using the results of a questionnaire survey and more general discussion of the professional
debate about EBM. Also looks at the impact of EBM on key constituents within health
care including purchasers, providers, patients and suppliers (particularly the pharmaceutical
and medical supplies industries).

(73)
WHAT physicians know
Tanenbaum, S J
New England Journal of Medicine, 21 Oct 1993 329(17) pp1268-71
A philosopher expresses reservations about EBM, urging doctors to defend clinical
reasoning based on experience and pathophysiological principles, and criticising the effects
of clinical epidemiology and health service research on clinical practice. For more from
Tanenbaum
See also: Getting there from here: evidential quandaries of the US outcomes movement,
by S Tanenbaum (Journal of Evaluation in Clinical Practice, 1995 1(2) pp97-103)

(74)
**EVIDENCE-based medicine means MDs must develop new skills, attitudes,
CMA conference told**
Rafuse, J
Canadian Medical Association Journal, 1 May 1994 150(9) pp1479-81
Reports on the 6th Annual Leadership Conference of the Canadian Medical Association
which was addressed by Gordon Guyatt, chair of the Evidence-Based Medicine Working
Group at McMaster University. The conference raised many cultural and practical
objections to EBM, with Dr Guyatt admitting that its practice was difficult because of a
number of factors. However, 'these challenges will be met'.

(75)
MEDICINE: a healing or a dying art?
Smith, B H; Taylor, R J
British Journal of General Practice, Apr 1996 46(405) pp249-51
The advent of EBM is serving to reinforce the popular belief that medicine is a science, and that its effective practice requires scientific proof. Argues that this is a narrow view which devalues medicine, and that the arts have important contributions to make to medical education and practice.

(76)
EVIDENCE-based medicine, in its place
Lancet, 23 Sep 1995 346(8978) p785
Criticises the EBM movement for stridency and for insisting that EBM is the only proper way to practise medicine. Although the movement is showing signs of greater caution, arguing that EBM builds on rather than disparages evidence gained from traditional clinical skills, it still threatens to alienate doctors who might otherwise support its principles. For letters in response from David Sackett and others
See also: Evidence-based medicine [letters] (Lancet, 28 Oct 1995 346(8983) pp1171-72)

(77)
THE PROTAGONISTS of 'evidence-based medicine': arrogant, seductive and controversial
Polychronis, A and others
Journal of Evaluation in Clinical Practice, Feb 1996 2(1) pp9-12, 30 references
Explores the case for describing the EBM movement in such critical terms, noting that the essence of the controversy lies in the alleged attempt of the EBM protagonists to make EBM 'the new medical orthodoxy', superior by definition to current modes of practice. Argues that the introduction of EBM into the purchasing process could produce systematic bias against the interests of patients in the sense that their freedom of choice may be reduced, and they may be denied access to effective treatments simply because no formal evidence exists to support them. In addition it offers NHS purchasers desperate to contain costs 'an apparently value-free, apolitical tool with which to excise large areas of health care expenditure'. For a related editorial, and for more from Polychronis and colleagues on clinical practice evaluation
See also: Evidence-based medicine: Reference? Dogma? Neologism?, by A Polychronis and others (Journal of Evaluation in Clinical Practice, Feb 1996 2(1) pp1-3, 12 references)
Central dimensions of clinical practice evaluation: efficiency, appropriateness and effectiveness: I, by D O'Neill and others (Journal of Evaluation in Clinical Practice, Feb 1996 2(1) pp13-27)
Central dimensions of clinical practice evaluation: efficiency, appropriateness and effectiveness: II, by A Miles and others (Journal of Evaluation in Clinical Practice, May 1996 2(2) pp131-52)

(78)
EVIDENCE based medicine [letters]
British Medical Journal, 22 Jul 1995 311(6999) pp257-59
Letters in response to an article by William Rosenberg and Anna Donald in *British Medical Journal*, 29 Apr 1995 310(6987) pp1122-26[16, Ch.1]. Many challenge the epidemiological basis of EBM.

(79)

LOOKING for the evidence in medicine

Taubes, G

Science, 5 Apr 1996 272(5258) pp22-24

Discusses the Cochrane Collaboration, including comment on the fears of critics that meta-analyses may mask errors and be inferior to judgement by top experts. For more on the problems of allowing for quality variations in meta-analysis

See also: Incorporating variations in the quality of individual randomized trials into meta-analysis, by A S Detsky and others (Journal of Clinical Epidemiology, Mar 1992 45(3) pp255-65)

(80)

WELCOME? To the brave new world of evidence based medicine

Carr-Hill, R

Social Science and Medicine, Dec 1995 41(11) pp1467-68

Examines the evolution of health services research, suggesting that the explosion of information is likely to produce more confusion than enlightenment. Researchers choose narrower and narrower topics of study, and repeat each other's work with pre-designed instruments, with the result that a mass of similar studies is cumulating around apparently easily answerable and very narrow themes. This sets the stage for systematic reviews which force healthcare into an even tighter straitjacket.

(81)

BRIDGES between health care research evidence and clinical practice

Haynes, R B and others

Journal of the American Medical Informatics Association, Nov/Dec 1995 2(6) pp342-50, 96 references

Looks at the development of EBM, with particular reference to the difficulties involved in getting research into practice. For more from Brian Haynes on this theme

See also: Applying evidence in clinical practice, by R B Haynes (Annals of the New York Academy of Sciences, 31 Dec 1993 703 pp210-24 + discussion pp224-25)

Loose connections between peer reviewed clinical journals and clinical practice, by R B Haynes (Annals of Internal Medicine, 1 Nov 1990 113(9) pp724-28)

(82)

WHERE medical science and human behaviour meet

Rees, J

British Medical Journal, 1 Apr 1995 310(6983) pp850-53

Looks at the likely future role of the consultant, including comment on the nature and impact of EBM. Argues that the idea that medical practice can be dictated by systematic evidence-based research is naive because it does not describe reality. Scientific evidence can never be more than a guide to medical practice which is underpinned not simply by a requirement to know the facts of a given situation, but also by the need to understand the personal and unique meaning that disease has for the individual patient.

(83)

QUANTIFICATION and the quest for medical certainty

Matthews, J R

Princeton University Press, 1996. 195pp

Looks at 'controversies over the use of statistical and probabilistic reasoning within medicine', arguing that both medical practitioners and researchers share an antipathy towards methods of quantification and statistical inference. The development of the double-blind clinical trial owes more to public demands for regulation than to the commitment of the medical profession.

(84)

CAN meta-analyses be trusted?
Thompson, S G; Pocock, S J
Lancet, 2 Nov 1991 338(8775) pp1127-30
Notes that the enthusiasm for meta-analysis of its proponents is not always shared by the wider medical community, and aims to promote constructive debate by looking at the conduct and interpretation of systematic reviews. Focuses particularly on statistical issues, especially heterogeneity between studies, and on the extrapolation of meta-analysis findings to clinical practice. Concludes that meta-analysis is not an exact statistical science able to provide definitive answers to clinical problems, but is more appropriately seen as a valuable descriptive technique. As such it can provide clear qualitative conclusions about broad treatment options, although its quantitative conclusions need to be interpreted with caution. For more on this theme, including comment on the failure of many Effective Health Care Bulletins to include the information on efficacy, cost-effectiveness and acceptability to patients that decision makers require
See also: Misleading meta-analysis, by M Egger and G Davey-Smith (British Medical Journal, 25 Mar 1995 310(6982) pp752-54)
Meta-analysis or best evidence synthesis, by H J Eysenck (Journal of Evaluation in Clinical Practice, Jan 1995 1(1) pp29-36)
Effective Health Care Bulletins: are they efficient? by D Torgerson and others (Quality in Health Care, Mar 1995 4(1) pp48-51)

(85)

REACHING the parts other methods cannot reach: an introduction to qualitative methods in health services research
Pope, C; Mays, N
British Medical Journal, 1 Jul 1995 311(6996) pp42-45, 19 references
Although qualitative research methods have a long history in the social sciences, they are frequently seen as antithetical to quantitative methods in medicine. Argues that they should be an essential component of health services research and not simply because they allow access to areas not amenable to quantitative methods. Qualitative description is a prerequisite of good quality quantitative research, particularly in areas that have received little investigation. For more on qualitative research
See also: Qualitative research methods in general practice and primary care, by N Britten and others (Family Practice, Mar 1995 12(1) pp104-14)
'Is my practice evidence-based?' should be answered in qualitative, as well as quantitative terms, by T Greenhalgh (British Medical Journal, 19 Oct 1996 313(7063) pp957-58)

(86)

EVIDENCE-BASED practice: the challenge for nursing, midwifery and health visiting
NHS Directorate for Wales, Cathays Park, Cardiff CF1 3NQ, Jun 1995. 19pp
Interest in evidence-based approaches to nursing is growing rapidly, with the first national conference on the subject held in late February 1997. For more recent comment on evidence-based approaches to nursing

See also: Interface between research and practice: some working models, by B Vaughan and M Edwards (King's Fund Centre, 11-13 Cavendish Square, London W1M 0AN, 1995. 54pp. Nursing Development Units)

Systematic search offers a sound evidence base, by S Holmes (Nursing Times, 24 Jan 1996 92(4) pp37-39)

Introducing an integrated nursing research programme, by C Shuldham and P Poole-Wilson (Nursing Standard, 24 Jan 1996 10(18) pp42-43)

Systematic reviews: keeping up with the research evidence, by R Dickson and V Entwistle (Nursing Standard, 31 Jan 1996 10(19) p32)

Using nursing initiatives to encourage the use of research, by D Mead (Nursing Standard, 31 Jan 1996 10(19) pp33-36)

Systematic reviews: how to use the information, by R Dickson and N Cullum (Nursing Standard, 7 Feb 1996 10(20) p32)

The process and importance of systematic reviews, by J Droogan and F Song (Nurse Researcher, Sep 1996 4(1) pp15-16)

Using meta-analysis to summarise evidence within systematic reviews, by J Greener and J Grimshaw (Nurse Researcher, Sep 1996 4(1) pp27-38)

(87)

EVIDENCE based medicine: implications for physiotherapy?

Partridge, C

Physiotherapy Research International, 1996 1(2) pp69-73

For more on evidence-based approaches and issues in relation to physiotherapy

See also: Evidence-based practice: how far have we come? by J Mead (Physiotherapy, Dec 1996 82(12) pp653-54)

Evidence-based practice: practice-based evidence, by M A Harrison (Physiotherapy Theory and Practice, Sep 1996 12(3) pp129-30)

(88)

PRIMARY care research: barriers and opportunities

Stange, K C

Journal of Family Practice, 1996 42(2) pp192-98

An American analysis which argues that a paucity of relevant scientific evidence limits the ability of general practitioners to practise EBM, and looks at how health care policy makers might respond. For more on EBM issues in general practice

See also: Evidence-based health care: the challenge for general practice, by A Hutchinson and others (In: Clinical effectiveness: from guidelines to cost-effective practice, edited by M Deighan and S Hitch. Earlybrave Publications Limited, PO Box 3165, Brentwood, Essex CM13 1TL, Oct 1995 pp49-52)

General practitioners and information: evidence-based practice explored, by F E Wood and others (In: Current perspectives in healthcare computing, edited by B Richards. BJHC: Weybridge, 1996 pp543-50)

(89)

GREY zones of clinical practice: some limits to evidence-based medicine

Naylor, C D

Lancet, 1 Apr 1995 345(8953) pp840-42, 21 references

The head of the Institute of Clinical Evaluative Sciences in Ontario argues that EBM offers little help in the many grey zones of medical practice where evidence on the pros and cons of competing clinical options is incomplete or contradictory. The grey zone is always likely to be large, and may be universal at the level of the individual patient, but the prudent use of the evaluative approaches which are characteristic of EBM can do much to enhance the art of medical practice.

(90)

FROM premise to principle: the impact of the gut hypothesis on the practice of critical care surgery

Marshall, J C; Girotti, M J

Canadian Journal of Surgery, Apr 1995 38(2) pp132-41, 82 references

Although clinical decision making is ideally based on evidence derived from well designed RCTs, in reality such evidence is rarely available to the surgeon caring for critically ill patients or those with multiple trauma. In these situations complex management decisions have to be made by more subjective means. Contrasts EBM, in which therapeutic decisions are made on the basis of systematic syntheses of RCTs, with inference-based medicine based on insights from studies in basic biology which are used to guide the practitioner's approach to groups of patients. Includes a case study of the impact of inference-based medicine on trauma and critical care surgery. For another analysis of the use of evidence-based approaches in intensive care, which argues that it does have a place in critical care

See also: Evidence-based critical care medicine: what is it and what can it do for us?, by D J Cook and others (Critical Care Medicine, Feb 1996 24(2) pp334-37)

(91)

HYPERTENSION and outcomes research: from clinical trials to clinical epidemiology

Psaty, B M and others

American Journal of Hypertension, 1996 9(2) pp178-83

Although RCTs usually provide the clearest evidence of effectiveness, they are often not done or not practicable. Thus, more than ten years after the introduction of calcium channel blockers and ACE inhibitors, there is still no clinical trial data available about their impact on major disease endpoints in patients with hypertension. Looks at the strengths and limitations of alternative sources of information: randomised trials including surrogate endpoints such as level of blood pressure or extent of carotid atherosclerosis; and observational studies including major disease endpoints.

(92)

COMPUTER searching of the medical literature: an evaluation of MEDLINE searching systems

Haynes, R B and others

Annals of Internal Medicine, Nov 1985 103(5) pp812-16

For more on the use of MEDLINE in clinical settings, the ability of relatively inexperienced users to access information effectively, and its value in identifying relevant information

See also: How to keep up with the medical literature V: access by personal computer to the medical literature, by R B Haynes and others (Annals of Internal Medicine, Nov 1986 105(5) pp810-16)

Bibliographic retrieval: a survey of individual users of MEDLINE, by K T Wallingford and others (MD Computing, May/Jun 1990 7(3) pp166-71)

How good are clinical MEDLINE searches? A comparative study of clinical end-user and librarian searches, by K A McKibbon and others (Computers and Biomedical Research, Dec 1990 23(6) pp583-93)

On-line access to MEDLINE in clinical settings: a study of use and usefulness, by R B Haynes and others (Annals of Internal Medicine, 1 Jan 1990 112(1) pp78-84)

A program to enhance clinical use of MEDLINE: a randomized controlled trial, by R B Haynes and others (Online Journal of Current Clinical Trials, 1993. Doc. No. 56)
Performances of 27 MEDLINE systems tested by searches with clinical questions, by R B Haynes and others (Journal of the American Medical Informatics Association, May/Jun 1994 1(3) pp285-95)
Developing optimal search strategies for detecting clinically sound studies in MEDLINE, by R B Haynes and others (Journal of the American Medical Informatics Association, Nov/Dec 1994 1(6) pp447-58)

(93)
ON the need for evidence-based general and family practice
Dawes, M G
Evidence-Based Medicine, Mar/Apr 1996 1(3) pp68-69
Looks briefly at the common arguments put forward against the practice of EBM in primary care, discounting each one in turn. For another brief analysis of the primary care EBM issues
See also: Evidence-based medicine and primary care, by J Bailey (IFMH Inform, Autumn/Winter 1996 7(3) pp5-8, 37 references)

(94)
PHARMACEUTICAL trials in general practice: the first 100 protocols: an audit by the clinical research ethics committee of the Royal College of General Practitioners
British Medical Journal, 16 Nov 1996 313(7067) pp1245-48
This study suggests that 'research sponsored by pharmaceutical companies and performed in general practice does not appear to generate a high level of scientifically valid, clinically relevant findings'. It surveys the outcome of the first 100 trials it considered, submitted by or on behalf of pharmacuetical companies between 1984 and 1989. Of the 82 trials approved, eight did not start. In those that did go ahead, 90% provided adequate data for the company to identify comparative efficacy of the drug. However, only 50% to 75% provided sufficient information to judge comparative tolerability, safety and acceptability. Only 19 of the studies have been formally published.

(95)
REVIEW of 39 years of randomised controlled trials in the British Journal of General Practice
Silagy, C A; Jewell, D
British Journal of General Practice, Aug 1994 44(385) pp359-63
An analysis of all trials published in the journal between 1953 and 1991 inclusive reveals 90 RCTs, 78% of which were undertaken in general practice. They covered pharmacological interventions (62%), non-pharmacological interventions (21%) and interventions related to the provision of an aspect of health service (16%). The quality of trials, measured by the degree to which bias was controlled, varied considerably. However, some imaginative solutions to the logistics of conducting RCTs in general practice were noted. For comment on the need for a register of RCTs in primary care
See also: Developing a register of randomised controlled trials in primary care, by C Silagy (British Medical Journal, 3 Apr 1993 306(6882) pp897-900)

(96)

EVIDENCE-based general practice: a critical reader

Ridsdale, L

BMJ Publishing: London, 1995. 182pp

Looks at the nature of general practice on the basis of real cases and published evidence. The final three chapters consist of a guide on how to appraise and evaluate scientific papers and other types of evidence in a critical, constructive and practical way. For further comment on the critical appraisal of medical literature by GPs, and for another primary care guide focusing specifically on cardiovascular disease

See also: READER: an acronym to aid critical reading by general practitioners, by D MacAuley (British Journal of General Practice, Feb 1994 44(379) pp83-85)

Critical appraisal of medical literature: an aid to rational decision making, by D MacAuley (Family Practice, Mar 1995 12(1) pp98-103, 6 references)

Prevention of cardiovascular disease: an evidence-based approach, edited by M Lawrence and others (Oxford University Press, 1996. 338pp. Oxford General Practice Series 33)

(97)

STORING a bibliographic database on your PC: a review of reference management software

Blumenthal, E Z; Gilad, R

New England Journal of Medicine, 22 Jul 1993 329(4) pp283-84

For more on the development of reference management software, including Pro-Cite

See also: Managing bibliographies with Pro-Cite, by T Hanson and A C Hanson (Aslib Proceedings, Feb 1989 41(2) pp75-83)

Personal bibliographic software programs: a comparative review, by P R Neal (BioScience, Jan 1993 43(1) pp44-51)

Personal computer software for handling references from CD-ROM and mainframe sources for scientific and medical reports, by R G Jones (British Medical Journal, 17 Jul 1993 307(6897) pp180-84)

(98)

EFFECTS of computer-based clinical decision support systems on clinician performance and patient outcome: a critical appraisal of research

Johnston, M E and others

Annals of Internal Medicine, 15 Jan 1994 120(2) pp135-42

Reviews the evidence from 28 controlled trials on the effects of computer-based clinical decision-support systems including computer-assisted dosing, computer-aided diagnosis, preventive care reminder systems, and computer-aided quality assurance for active medical care. The results show strong evidence that some systems can improve clinical performance, though more studies are needed to assess cost-effectiveness, especially in the context of patient outcome.

(99)

THE CRITICALLY appraised topic: a practical approach to learning critical appraisal

Sauve, S and others

Annals of the Royal Society of Physicians and Surgeons of Canada, Oct 1995 28(7) pp396-98

Describes the paper-based system developed by General Internal Medicine Fellows at McMaster University for generating good quality critical appraisals in a standardised format. For details of the CATmaker software under development by the UK Centre for

Evidence-Based Medicine as an electronic version of this system, contact Douglas Badenoch, Programme Manager for Education and Communication, Centre for Evidence-Based Medicine, Level 5, John Radcliffe Hospital, Headley Way, Headington, Oxford OX3 9DU.

(100)
PRODIGY official in UK
SCRIP, 10 Nov 1995 (2076) p4
Brief news item on the rational prescribing project, PRODIGY, which is designed to give GPs quick access to prescribing alternatives for the bulk of the conditions they encounter. For the Audit Commission's views on rational prescribing in general practice, the interim report on PRODIGY, and additional comment on the worries of pharmaceutical companies and patients' groups about its implications
See also: A prescription for improvement: towards more rational prescribing in general practice, by Audit Commission (HMSO (now Stationery Office), 1994. 104pp. Health and Personal Social Services Reports 1 1994. 27pp executive summary also available)
PRODIGY interim report, by I Purves (University of Newcastle-upon-Tyne, Sowerby Unit for Primary Care Informatics, 1996. 13pp. PRODIGY Publication 24)
DoH spends £1.5 million on PRODIGY (SCRIP, 12 Jan 1996 (2093) p4)

(101)
WHAT the doctor ordered: a study of GP fundholders in England and Wales
Audit Commission
HMSO (now Stationery Office), May 1996. 136pp
Finds that many fundholders are 'under-performing or making purchasing decisions which represent less than the best possible value for money'. Few are making full use of clinical effectiveness information, in part because of demands from patients for complementary therapies which have not been proven to be effective. For the results of a recent survey of the use of research evidence by Scottish doctors (GPs and senior medical managers) which also shows major barriers to the implementation of an evidence-based approach despite general support for the concept
See also: Decision-making by health purchasing organisations in Scotland: the role and influence of evidence from the research literature, by J Farmer and D Williams (Journal of Information Science, 1997 23(1) pp59-72, 27 references)

(102)
ACTING on the evidence: a review of clinical effectiveness: sources of information, dissemination and implementation
Appleby, J and others
National Association of Health Authorities and Trusts, Birmingham Research Park, Vincent Drive, Birmingham B15 2SQ, 1995. 38pp, references (NAHAT Research Paper 17)
Examines the fast-growing interest in evidence-based health care within the NHS, looking at both the potential of EBM and the problems involved in implementation. The latest assessment of progress by the research team from the University of Birmingham which produced the report suggests that many performance incentives within the NHS still continue to stress economy and efficiency rather than clinical effectiveness. As a result, needless or ineffective procedures continue to be carried out, while effective ones are ignored. For journal comment by the authors of this influential report, and for their most recent analysis

See also: Given in evidence, by K Walshe and others (Health Service Journal, 29 Jun 1995 105(5459) pp28-29)

Evidence based healthcare: brave new world? by K Walshe (Clinical Risk Report, 1996 2(4) pp16-18)

Acting on the evidence: progress in the NHS, by K Walshe and C Ham (NHS Confederation, Birmingham Research Park, Vincent Drive, Birmingham B15 2SQ, Apr 1997. 37pp)

From evidence into action, by K Walshe (Health Director, Apr 1997 pp8-10)

Who's acting on the evidence? by K Walshe and C Ham (Health Service Journal, 3 Apr 1997 107(5547) pp22-25)

Note: NAHAT has been renamed the NHS Confederation. Tel: 0121 471 4444; Fax 0121 414 1120; URL **http://www.nahat.net**

(103)
EVIDENCE based contracting for health care
Peach, E and others
Nuffield Institute of Health, University of Leeds, and Yorkshire Collaborating Centre for Health Services Research, Dec 1995. 33pp

(104)
METHODS to promote the implementation of findings in the NHS: priorities for evaluation
Department of Health, Richmond House, 79 Whitehall, London SW1A 2NS, Oct 1995. 65pp
A report from one of the *ad hoc* expert working groups serving the Central Research and Development Committee set up by the Peckham Report. Chaired by Professor Andrew Haines, R&D Director of the North Thames health authority, it looks in detail at a wide range of possible interventions, and identifies 20 priority areas for further research. For further comment from Andrew Haines
See also: Implementing findings of research, by A Haines and R Jones (British Medical Journal, 4 Jun 1994 308(6942) pp1488-92)

(105)
IMPLEMENTING findings of medical research: the Cochrane Collaboration on Effective Professional Practice
Freemantle, N and others
Quality in Health Care, Mar 1995 4(1) pp45-47, 18 references
The focus of this particular Cochrane Collaboration is on reviews of trials and interventions designed to improve professional practice through, for example, various forms of dissemination, continuing education and quality assurance. It also examines reviews of trials of financial and organisational interventions designed to help health care providers deliver services more effectively.
Note: This Collaboration, under the leadership of Jeremy Grimshaw as co-ordinating editor, moved from York in May 1997 and is now based at the Health Services Research Unit, Department of Public Health, University of Aberdeen, Drew Kay Wing, Polwarth Building, Foresterhill, Aberdeen AB25 2ZD (Tel: 01224 681818 ext. 51100; Fax: 01224 663087; e-mail ccepp@abdn.ac.uk)

(106)
R&D priorities in relation to the interface between primary and secondary care: report to the NHS Central Research and Development Committee
Department of Health, Richmond House, 79 Whitehall, London SW1A 2NS, Mar 1994. 72pp

(107)
NO magic bullets: a systematic review of 102 trials of interventions to improve professional practice
Oxman, A D and others
Canadian Medical Association, 15 Nov 1995 153(10) pp1423-31, 131 references
The analysis shows that dissemination–only strategies such as conferences or the mailing of unsolicited information have very little impact when used alone. The outcome of more complex interventions such as the use of outreach visits or local opinion leaders ranged from highly effective to totally ineffective, but was most usually moderately effective, resulting in reductions of 20% to 50% in the incidence of inappropriate performance. This study was supported by Health Canada, the North Thames Regional Health Authority R&D Programme and the UK Cochrane Centre. For Andrew Oxman's original report, and for more on the research-into-practice problem
See also: No magic bullets: a systematic review of 102 trials of interventions to help health care professionals deliver services more effectively or efficiently, by A D Oxman (North Thames Regional Health Authority, 40 Eastbourne Terrace, London W2 3QR, Mar 1994. 49pp)
From dissemination to use: management and organisational barriers to the application of health services research findings, by P Williamson (Health Bulletin, 1992 50(1) pp78-86)
Dissemination of effectiveness and outcomes research, by D Kanouse and others (Health Policy, 1995 34(3) pp167-92)

(108)
THE DISSEMINATION of R&D information
Watt, I
IFHM Inform, Spring 1996 7(1) pp1-4, 6 references
Looks briefly at the research–to–practice problem, noting that simple dissemination strategies rarely work. Recommends a mix of traditional dissemination activities with implementation initiatives designed to get information into practice, focusing on the importance of locally based action.

4.2 Teaching and learning EBM

The development of an evidence-based health care system demands both new skills and changes in attitudes on the part of practitioners if the scepticism reflected in the professional debate about EBM is to be overcome and the practical difficulties of implementation addressed. In particular there needs to be a commitment to continuous learning and information awareness throughout working life. Education and in-service training are, therefore, of prime importance in ensuring that effective links between research and practice are made. However, it has been frequently argued that conventional continuing medical education fails to modify clinical performance or the health outcomes of patients although it may lead to a significant increase in practitioner knowledge[109].

Clearly a new approach is needed to stem the continuing decline in professional competence following completion of formal training. Much of the impetus for development in this area has come from David Sackett and his colleagues through their pioneering work at McMaster University in teaching critical appraisal skills to both undergraduates and qualified doctors wishing to pursue the EBM model. Participants in McMaster's Internal Medicine Residency Program for qualified doctors are carefully selected on the basis of their enthusiasm and commitment to EBM principles, on the understanding that the wider development of EBM is crucially dependent on the influence of respected role models[29,Ch.1].

The course of weekly, half-day sessions begins with a review of the rules of evidence relating to scientific articles concerned with therapy, diagnosis, prognosis and research overviews. In subsequent sessions, participants develop their critical appraisal skills by examining original articles in the context of real clinical cases. Facilities for computerised literature searching are available on the ward, and training is provided for those who lack searching skills. Finally, participants are taught techniques of physical examination that will give the optimal return in terms of accuracy and reproducibility. The McMaster team is fully aware of the scepticism – and occasional fear – with which many doctors view its work, and understand the need to reduce the threatening aspects of EBM by setting modest and achievable goals, and building critical appraisal skills through experience. Careful selection of initial EBM exercises is also important to ensure that high quality evidence is there to be retrieved by novice practitioners.

Although it may be impossible to prove conclusively that EBM provides a better outcome for the patient than the traditional model of practice, it is (in the view of its proponents) a reasonable assumption. Moreover, they believe that their experience proves that EBM skills can be taught effectively to both undergraduates and qualified practitioners[110]. A trial of the impact of McMaster's undergraduate curriculum shows that EBM-trained students are better able to generate and defend diagnostic and management decisions than those receiving a conventional medical education[111]. McMaster graduates also retain a high knowledge of clinically important advances, even 15 years after qualifying, while their peers show the usual progressive deterioration in this area of competence[112].

McMaster University has not confined its EBM educational activities to those able to attend in person as undergraduates or postgraduates. In the early 1980s it also published a series of 'readers' guides' for clinicians to follow when reading the medical literature. These were heavily requested, reprinted in several languages and modified for use by the general public, and have been updated in the *Users' guides to the medical literature* published in the *Journal of the American Medical Association* between 1993 and 1995[113]. David Sackett and his colleagues have also recently published a 250pp guide to the practice and teaching of EBM which has been well received by the British medical press, though with continuing reservations about the realism of its aims[114]. One reviewer describes feeling 'rather mauled by this book. All you need to practise and teach evidence based medicine is in this book (except for the second 24 hours in every day)'.

Training initiatives
The McMaster approach has been extensively adopted in other countries including the UK, and there is a variety of training initiatives in progress or under discussion including the establishment of a Centre for Evidence-Based Nursing. Existing programmes include:

- **THE CENTRE FOR EVIDENCE-BASED CHILD HEALTH**
  Department of Epidemiology and Biostatistics
  Institute of Child Health
  30 Guildford Street
  London WC1N 1EH

  Tel: 0171 242 9789
  Email: r.gilbert@ich.ucl.ac.uk
  URL: **http://www.ich.bpmf.ac.uk/ebm/ebm.htm**

  The newly opened Centre for Evidence-Based Child Health headed by Dr Ruth Gilbert and Dr Stuart Logan is part of a national network of centres for evidence-based health care which is focusing on educational programmes for health professionals. These include introductory seminars to raise EBM awareness; short courses in basic literature searching and critical appraisal skills; MSc modules in clinical epidemiology; and workplace-based workshops focused around specific clinical problems identified by the group as important. The Centre also plans to offer 3-12 month training secondments for paediatricians, nurses, GPs, health care purchasers and others involved in child health to implement research findings in their particular service. It is a joint participant in the Systematic Reviews Training Unit (see below).

- **THE CENTRE FOR EVIDENCE-BASED DENTISTRY**
  Institute of Health Sciences
  Old Road
  Headington
  Oxford OX3 7LF

  Tel: 0118 982 2860

  A new initiative set up following a workshop on evidence-based dentistry in December 1994[115]. The aims of the Centre are to promote the teaching, learning, practice and evaluation of evidence-based dentistry throughout the UK. It has already held critical appraisal workshops, run initially with the help of the Critical Skills Appraisal Programme (see below), and is developing 'find the evidence workshops' in association with CASP and the Health Libraries Information Network. Teaching materials are being developed with the aim of producing an evidence-based workbook for dentists.

- **THE CENTRE FOR EVIDENCE-BASED MEDICINE**
  Level 5
  John Radcliffe Hospital
  Headley Way
  Headington
  Oxford OX3 9DU

  Tel: 01865 221320
  Fax: 01865 222901
  Email: David.Sackett@ndm.ox.ac.uk
  URL: **http://cebm.jr2.ox.ac.uk**

  Members of the Centre for Evidence-Based Medicine act as EBM pioneers by practising EBM in their everyday clinical work, and by undertaking visiting professorships at other institutions in the UK and Europe. In addition, the Centre runs workshops on how to practise EBM; how to teach EBM; and how to introduce and evaluate EBM teaching programmes. The workshop programme is based on David Sackett's extensive experience in running such events elsewhere, and a top priority will be the training of additional EBM tutors. Other educational initiatives include collaborating in the introduction and expansion of EBM teaching in other clinical schools and postgraduate training programmes throughout the UK, and working with national professional organisations involved in continuing professional development. The Centre is also involved in the promotion of the teaching, learning, practice and evaluation of evidence-based approaches to non-medical health care.

- **THE CRITICAL APPRAISAL SKILLS PROGRAMME (CASP)**
  Oxford CASP Office
  PO Box 777
  Oxford OX3 7LF

  Fax: 01865 226959
  Email: casp@cix.compulink.co.uk
  URL: **http://fester.his.path.cam.ac.uk/phealth/casphome.htm**

  CASP is based in the Anglia and Oxford region but includes links to other parts of the country. Its aim is 'to help health service decision makers develop skills in the critical appraisal of evidence about effectiveness, in order to promote the delivery of evidence-based health care. At the heart of CASP's work is a cascade of half-day workshops. They introduce participants to the key skills needed to find and make sense of evidence to support health service decisions'. The workshops focus on the critical appraisal of systematic reviews, RCTs and economic evaluations using checklists adapted from the *Users' guides* published in the *Journal of the American Medical Association*, and are designed principally for GPs, health service managers, public health specialists and others involved in purchasing. CASP thus complements the work of the Centre for Evidence-Based Medicine which works directly with clinicians providing care to patients[116].

- **EVIDENCE BASED HEALTH DISCUSSION LIST**

  Email: mailbase@mailbase.ac.uk
  URL: **http://www.mailbase.ac.uk/lists-a-e/evidence-based-health**

  This UK mailing list is designed for teachers and practitioners in health-related fields to announce meetings and courses, stimulate discussion, air controversies and assist the implementation of evidence-based health care. Subscribers should send the email message 'join evidence-based health A N Other' (substitute your own name) to the address above.

- **NORTH THAMES RESEARCH APPRAISAL GROUP**
  Department of Epidemiology and Medical Statistics
  London Hospital Medical College at QMW
  London E1 4NS

  Tel: 0171 982 6328
  Fax: 0171 982 6396

The Group runs workshops teaching critical appraisal skills to health professionals in the North Thames Region. Courses are delivered in a variety of formats depending on the needs of particular groups and cover the critical appraisal of primary studies as well as systematic reviews and economic analyses.

Other developments in the North Thames Region can be accessed via the home page of the North Thames Regional Library and Information Unit at **http://www.nthames-health.tpmde.ac.uk/ntrl** which includes information on R&D and education/training developments in the region. It also provides links to other health care (including EBM) resources. The Unit is based at Thames Postgraduate Medical and Dental Education, 33 Millman Street, London WC1N 3EJ (Tel: 0171 405 5667; Fax: 0171 405 5668).

- **SHEFFIELD CENTRE FOR HEALTH AND RELATED RESEARCH**
  Regent Court
  30 Regent Street
  Sheffield S1 4DA

  Tel: 0114 276 8555 ext. 5454/5455
  Fax: 0114 272 4095
  Email: scharrlib@sheffield.ac.uk
  URL: **http://www.shef.ac.uk/uni/academic/R-Z/scharr/**

SCHARR undertakes teaching, training, research and consultancy work into all aspects of health service research and technology assessment. Educational initiatives include taught MSc courses in health economics and management, and health services research and technology assessment; and higher research degrees. It also provides an online medical informatics course including tutorials and practice exercises.

SCHARR is currently working with Sheffield University Department of Information Studies to develop a computer-based reference service capable of increasing the range of EBM information queries that can be handled by machine. The AuRACLE (Automated Retrieval Assistant for CLinically-relevant Evidence) project will develop an optimal, structured request form for literature searches in health care libraries; develop an intelligent interface to health care databases such as MEDLINE to maximise the efficiency of end-user searching; and incorporate the search skills of library and information staff into the machine system. For more details contact Jane Ralph or Alan O'Rourke of the Department of Information Studies, Regent Court, 211 Portobello Street, Sheffield S1 4DP (Email: j.l.ralph@shef.ac.uk, or a.j.rourke@shef.ac.uk).

- **SYSTEMATIC REVIEWS TRAINING UNIT**
  Department of Epidemiology and Biostatistics
  Institute of Child Health
  30 Guildford Street
  London WC1N 1EH

Tel: 0171 242 9789
Fax: 0171 813 8233
Email: slogan@ich.ucl.ac.uk
URL: **http://www.ich.bpmf.ac.uk/ebm/srtu.htm**

This initiative has been set up with funding for three years by the North and South Thames Regional Research and Development Programme, and is a joint project of the Institute of Child Health, the Royal Free Hospital School of Medicine and University College London. It is supported by a Collaborative Reviews Group whose members are specialists in a variety of medical disciplines, and its main aim is to provide training for health professionals in the conduct, dissemination and promotion of systematic reviews.

- **UNIVERSITY OF OXFORD DEPARTMENT FOR CONTINUING EDUCATION AND THE FACULTY OF CLINICAL MEDICINE: CONTINUING PROFESSIONAL DEVELOPMENT PROGRAMME**
  1 Wellington Square
  Oxford OX1 2JA

  Tel: 01865 280347
  Fax: 01865 270386
  Email: elaine.welsh@conted.ox.ac.uk or venetia.hill-perkins@conted.ox.ac.uk

  The Oxford Master's Programme in Evidence-Based Health Care is a part time integrated programme of professional development in evidence-based health care, comprising a postgraduate certificate, postgraduate diploma and MSc.

- **WORKSHOP ON HOW TO TEACH EVIDENCE BASED MEDICINE**
  McMaster University Department of Clinical Epidemiology and Biostatistics
  Health Information Research Unit
  1200 Main Street West
  Hamilton
  Ontario

  URL: **http://hiru.hirunet.mcmaster.ca/ebm/workshop/**

  The Department of Clinical Epidemiology and Biostatistics has developed workshops to help other medical schools develop undergraduate and postgraduate training programmes along the lines of the successful McMaster model. As such, they help participants to learn how to teach critical appraisal and EBM (and develop their own skills) and generate educational packages for local use. The workshops are based on packages of readings dealing with EBM and critical appraisal issues in relation to therapy, diagnosis, prognosis, harm, overviews and economic analysis. Some of this material is available via the Internet and can be downloaded to support local critical appraisal training initiatives in the UK.

Undergraduate education

Major changes have also been introduced in undergraduate medical education in the UK in response to the growing deficiencies of the traditional model in the face of external changes such as the increasing interest in public (as well as individual) health, a shift in the balance between hospital-based and community care, the impact of an ageing population, the quickening pace of medical advance (and its associated ethical and moral dilemmas), and the increase in public understanding of disease and disability. In addition, the traditional model of medical education involved considerable factual overload on students, and the inevitable blunting of their enthusiasm and ability to think independently. A report from the King's Fund in 1991 on the crisis in undergraduate medical education was particularly influential in prompting reforms which had been called for over many years[117].

Tomorrow's doctors published by the General Medical Council in 1993 seeks to reduce the burden of factual information imposed on students (though it remains substantial), and develop habits of curiosity and self-directed learning which are fully in tune with the principles of EBM[118]. The curriculum theme of 'Finding out: research and experiment' should 'permeate all aspects of the course'. 'Learning through curiosity, the exploration of knowledge, and the critical evaluation of evidence should be promoted and should ensure a capacity for self-education; the undergraduate course should be seen as the first stage in the continuum of medical education that extends throughout professional life'.

A very similar approach is taken by the British Medical Association in its 1995 working party report on medical education which deals with all stages of the medical career[119]. It stresses the importance of good basic understanding of the mechanisms of the body but notes that 'medical knowledge is constantly and rapidly changing'. Thus the key factor in maintaining and improving skills is the instilling of 'a curiosity and a desire to know more, along with the ability to find the information they want or need on a given occasion'. Once these attitudes and abilities are inbuilt, 'continuing medical education and professional development become second nature' and an integral part of the doctor's life 'from entry into medical school to retirement'.

(109)
EVIDENCE for the effectiveness of CME: a review of 50 randomized controlled trials
Davis, D A and others
Journal of the American Medical Association, 2 Sep 1992 268(9) pp1111-17
Outlines the debate over the effectiveness of continuing medical education, noting that studies suggest a positive impact on clinical competency (the doctor's ability to perform in test situations) but less strong evidence for changes in clinical practice or patient outcome. The analysis of RCTs shows that CME interventions using practice-enabling or reinforcing strategies consistently improve clinical performance and, in some instances, health care outcomes. For later comment
See also: Changing physician performance: a systematic review of the effect of continuing medical education strategies, by D A Davis and others (Journal of the American Medical Association, 6 Sep 1995 274(9) pp700-05)

(110)
TEACHING residents to read the medical literature: a controlled trial of a curriculum in critical appraisal/clinical epidemiology
Kitchens, J M; Pfeifer, M P
Journal of General Internal Medicine, Sep/Oct 1989 4(5) pp384-87
The trial involved allocating a group of residents to eight ambulatory care clinics for a half day each week. Four groups (Group A) were exposed to pre-clinic conferences on critical appraisal skills, while the other four (Group B) had conferences on aspects of ambulatory

care medicine. At the end of Phase One of the trial, the groups were given a test of basic knowledge of clinical epidemiology. The curriculum was then modified, and the groups switched round with Group B now exposed to critical appraisal skills. At a second test Group B performed significantly better than at the first test, while Group A had not improved.

(111)
A CONTROLLED trial of teaching critical appraisal of the clinical literature to medical students
Bennett, K J and others
Journal of the American Medical Association, 8 May 1987 257(18) pp2451-54
Describes a controlled trial in which two groups of final year undergraduates, and their teachers, were offered a short course in the critical appraisal of the medical literature, and compared with two groups that did not. The experimental groups showed statistically and 'clinically' significant improvements in their appraisal skills relating to a diagnostic test and a treatment exercise. Control students' scores deteriorated for both tests.

(112)
EFFECT of problem-based, self-directed undergraduate education on life-long learning
Shin, J H and others
Canadian Medical Association Journal, 15 Mar 1993 148(6) pp969-76
Presents the results of an analytic survey to test how well graduates of the McMaster University medical school (self-directed, problem-based curriculum) compare with graduates of Toronto University (traditional curriculum) in respect of keeping up to date with clinical practice guidelines. The test situation focused on management of hypertension among graduates working in general practice. The McMaster graduates proved to be significantly more up to date in their knowledge of recommended blood pressures for treatment, and of successful approaches to enhance compliance with treatment.

(113)
USERS' guides to the medical literature
Evidence-Based Medicine Working Group
Journal of the American Medical Association (a complete list with full bibliographical references is also available via **http://www.soton.ac.uk/~swhclu/schebm.htm** while full text of some of the guides can be accessed via **http://hiru.hirunet.mcmaster.ca/ ebm/userguid/default.htm**)
Guides published since November 1993 are as follows:

I:	How to get started (3 Nov 1993 270(17) pp2093-95)
IIA:	How to use an article about therapy or prevention A: are the results of the study valid? (1 Dec 1993 270(21) pp2598-601)
IIB:	How to use an article about therapy or prevention B: what were the results and will they help me in caring for my patients? (5 Jan 1994 271(1) pp59-63)
IIIA:	How to use an article about a diagnostic test A: are the results of the study valid? (2 Feb 1994 271(5) pp389-91)
IIIB:	How to use an article about a diagnostic test B: what are the results and will they help me in caring for my patients? (2 Mar 1994 271(9) pp703-07)
IV:	How to use an article about harm (25 May 1994 271(20) pp1615-19)
V:	How to use an article about prognosis (20 Jul 1994 272(3) pp234-37)

VI: How to use an overview (2 Nov 1994 272(17) pp1367-71) [Comment in JAMA, 19 Jul 1995 274(3) pp217-18]

VIIA: How to use a clinical decision analysis A: are the results of the study valid? (26 Apr 1995 273(16) pp1292-95)

VIIB: How to use a clinical decision analysis B: what are the results and will they help me in caring for my patients? (24/31 May 1995 273(20) pp1610-13)

VIIIA: How to use clinical practice guidelines A: are the recommendations valid? (16 Aug 1995 274(7) pp570-74)

VIIIB: How to use clinical practice guidelines B: what are the results and will they help me in caring for my patients? (22/29 Nov 1995 274(20) pp1630-32)

IX: A method for grading health care recommendations (13 Dec 1995 274(22) pp1800-04)

(114)
EVIDENCE based medicine: how to practice and teach EBM
Sackett, D L and others

Churchill Livingstone, Robert Stevenson House, 1-3 Baxter's Place, Leith Walk, Edinburgh EH1 3AF, 1996. 250pp (Available through the British Medical Association Bookshop to BMA members)

An introduction on the need for evidence-based medicine is followed by advice on how to ask a clinical question; search for the best evidence; critically appraise it; apply it to the care of patients; and evaluate one's use of evidence. For reviews

See also: Evidence-based medicine: how to practice and teach EBM, by D Sharp (Lancet, 9 Nov 1996 348(9037) p1297)

Evidence-based medicine: how to practice and teach EBM, by K Jones (British Medical Journal, 30 Nov 1996 313(7069) p1410)

(115)
EVIDENCE-based dentistry
Richards, D; Lawrence, A

British Dental Journal, 7 Oct 1995 179(7) pp270-73, 18 references

Reviews the problems of introducing evidence-based dentistry including lack of clinically relevant evidence (many widely read dental publications are not subject to peer review) and information overload. Looks briefly at what constitutes good evidence, how to find it, and how to make sense of it. Concludes with brief comment on initiatives in evidence-based dentistry. For more comment

See also: Developing evidence-based dentistry, by D Richards (Primary Dental Care, Mar 1996 3(1) pp4-5)

A Centre for Evidence-Based Dentistry, by D Richards (Journal of Clinical Effectiveness, 1996 1(2) p70)

(116)
CRITICAL Appraisal Skills Programme: making sense of evidence about clinical effectiveness: orientation guide 1995
Critical Appraisal Skills Programme, PO Box 777, Oxford OX3 7LF, 1995. 14pp

Outlines the CASP approach and includes reprints of articles on EBM from *New Scientist* (21 Jan 1995 145(1961) pp14-15) and the *Journal of the American Medical Association.*

(117)
CRITICAL thinking: the future of undergraduate medical education
Towle, A

King's Fund Centre, 11-13 Cavendish Square, London W1M 0AN, 1991. 38pp

(118)
TOMORROW'S doctors
General Medical Council Education Committee
General Medical Council, 44 Hallam Street, London W1N 6AE, Dec 1993. 28pp
Looks at the external and internal factors demanding the reform of undergraduate medical education, and describes the objectives to be achieved in terms of knowledge, skills and attitudes. Also identifies curriculum themes and discusses the way in which undergraduate education should be delivered, favouring a self-directed, problem-based, modular approach. The Department of Health has subsequently provided special funding to ensure the smooth introduction of curricular changes. For a useful summary of the radical changes in undergraduate medical and dental education in recent years
See also: Undergraduate medical and dental education: fourth report of the Steering Group (Steering Group on Undergraduate Medical and Dental Education and Research, Mar 1996. 49pp)

(119)
REPORT of the working party on medical education
British Medical Association, BMA House, Tavistock Square, London WC1H 9JP, 1995. 31pp
The working party was established in response to increasing calls for the reform of undergraduate medical education from the General Medical Council, King's Fund and others. However, it covers the entire spectrum of medical education from entry to medical school to retirement, stressing the importance of instilling a desire to learn and the ability to find and use new evidence when this is needed.

4.3 Audit and guidelines

The educational initiatives discussed in the last section are designed to promote an evaluative culture from within by teaching new skills and developing new attitudes. These are by their nature slowly developing processes which depend on the movement of young EBM-trained doctors into the health service, and the gradual spreading of the message by EBM champions already in positions of influence. These include sympathetic fundholding GPs, hospital clinical and medical directors, and public health physicians involved in commissioning. However, the NHS is also concerned to promote change through managerial innovations with a more immediate impact. Included among these are clinical audit, introduced by the 1989 White Paper *Working for patients*[36,Ch.2], and clinical guidelines.

Clinical audit is the process by which doctors, nurses and other health care professionals systematically review and, where necessary, make changes to the care and treatment they provide to patients[120]. It has been practised by some clinicians for many years, and there is a substantial clinical audit literature emanating from the Royal Colleges and other bodies[121]. The reports of CASPE (Clinical Accountability Service Planning and Accountability) Research provide a useful review of some of this activity[122]. However, audit is by no means universal and, between 1989-90 and 1993-94 the NHS Executive provided some £218 million for the introduction and development of projects. By 1993-94 more than 20,000 clinical audit projects had been carried out in hospital and community health services, and the NHS Executive now intends that audit should become an integral part of NHS practice for all health care professionals[123]. Health authorities have had responsibility for clinical audit since April 1996, although they will be monitored by the NHS Executive regional offices, and funding for audit activities will no longer be separately identified[124].

An analysis of clinical audit activity in three English health regions by the National Audit Office in late 1995 claims that a third of all projects led to measurable change in clinical care practice and/or organisation, with 25% of these involving the use or development of clinical guidelines[125]. However, the principal impact appears to be on the health care delivery process, teamworking and other management areas, and relatively few changes appear to have a direct impact on the quality of care or care outcomes. The NHS Executive is clear that there needs to be greater collaboration between the NHS and the professional bodies if clinical audit is to deliver maximum benefit, and has established a National Centre for Clinical Audit under contract to a partnership of the British Medical Association and the Royal College of Nursing. The Centre, based at the BMA (BMA House, Tavistock Square, London WC1H 9JP. Tel: 0171 383 6451; Fax: 0171 383 6373), promotes best practice gained from local initiatives and multi-professional audits, and is overseen by the Clinical Outcomes Group (NHS Executive, HCD-PH1B Room 408, Wellington House, 135-155 Waterloo Road, London SE1 8UG. Tel: 0171 972 4926; Fax: 0171 972 4673).

Specialist clinical audit is also an important element of the monitoring side of the NHS effectiveness 'triangle' described in *Promoting clinical effectiveness*[62,Ch.2]. Monitoring is centred on the NHS Executive's Information Technology and Management Strategy and includes the development of comparative data sets based on routinely collected information such as the Public Health Common Data Set; data sets based on detailed feasibility studies such as the Health of the Nation NHS Performance Measures; and national audits undertaken by the Clinical Standards Advisory Group, an independent, multi-disciplinary professional body looking in detail at standards of care across the NHS[126]. In addition, a range of national audits (some confidential) has been carried out by expert groups into areas such as hospital infection control, stillbirths and deaths in infancy, major trauma services, and intensive care[127].

Clinical guidelines
Clinical audit is seen by the NHS Executive as 'the framework within which clinical guidelines, needs assessment, evidence of effectiveness and information on cost effectiveness can all be brought together to provide a focus which can improve the quality of clinical care' under the aegis of local Medical Audit Advisory Groups[128]. Audits may be carried out in association with existing guidelines on the clinical management of particular groups of illnesses but, in a significant number of cases, these are of inadequate quality. Some guidelines, for example, do not show an explicit link between the evidence on which they are based and the recommendations they contain, while many are constructed through a committee procedure and are based on informal consensus rather than rigorous scientific evidence (which may, in many cases, be lacking)[129].

Sound information on cost effectiveness is in even shorter supply, although skills in this area are developing[130]. The Department of Health commissioned the University of York's Centre for Health Economics to produce a *Register of cost effectiveness studies* in 1994, listing some 150 economic evaluations of healthcare services, treatments, procedures and other medical technologies[131]. This is now maintained and updated by the NHS Centre for

Reviews and Dissemination as the NHS Economic Evaluation Database which, as noted in Chapter 3, includes structured abstracts of published economic evaluations of health care interventions, with an assessment of quality and the practical implications for the NHS.

The Institute of Medicine in the USA identifies cost effectiveness information as a desirable attribute of a clinical guideline[132], and this has been defined in the UK (in *Effective Health Care Bulletin* 8) as 'guidelines which lead to improvements in health at acceptable costs'[133]. Some commentators argue that cost effectiveness information is not merely a desirable attribute of a clinical guideline but an essential component; appropriate recommendations must be based on cost per unit of health gain as well as outcome probabilities and expected health gain[134]. This may well be the ideal given the fact that limited resources mean there will always be opportunity costs associated with the decision to invest in this or that health care intervention. However, until economic evaluation becomes a routine component of health care R&D, it will be difficult to achieve[135].

In response to the acknowledged deficiencies of many existing clinical guidelines, the NHS Executive, on the advice of the Clinical Outcomes Group, is identifying and recommending good quality guidelines based on RCTs, 'other robust experimental or observational studies', or more limited evidence that is endorsed by 'respected authorities'[136]. The definition of what constitutes quality in evidence-based guidelines has been developed in association with the professional bodies although these will remain free to issue their own guidelines if they so wish[137]. In areas which lack good quality guidelines, the activities of the Cochrane Collaboration and other bodies involved in systematic review of the medical literature will clearly play a major role. The clinical audit process will also provide important input into the development of guidelines at local or national level which the COG intends to be used by all those involved in health care decision making.

These include not just clinicians who need to keep up to date with properly evaluated new information on the clinical and cost-effectiveness of interventions in particular areas. Other audiences for clinical guidelines include national policy makers who increasingly view guidelines as a tool for helping to determine where resources are best allocated, and where they may be safely reduced or withdrawn. One commentator describes this as an approach to uncertainty which 'attempts to uncover the best available evidence, the stakeholders' interpretations of the evidence and the underlying values of the stakeholders'[138]. Commissioning agencies including health authorities, groups of doctors, insurance agencies and others can also make use of guidelines in deciding on the most effective use of limited resources. Thus increasing numbers of bodies have used guidelines to restrict access to certain types of treatment such as D&C which evidence shows to have limited value. Finally, guidelines – suitably repackaged – are also seen as playing a potentially major role in helping to inform and empower the patient.

Implementation and evaluation

Despite the attention devoted to clinical guidelines by the NHS and health services in other countries, relatively little attention has been paid to ensuring implementation or evaluating impact. For example, a preliminary survey of 55 organisations involved in the 1994 Canadian Clinical Practice Guidelines Network Workshop showed that more than 60% failed to make any active attempt to ensure the implementation of the guidelines they disseminated to practitioners[139]. Dissemination/implementation activities were dominated by direct mailing, publication in journals and newsletters, and the organisation of conferences or workshops, with much less attention paid to more active strategies involving training, outreach vists, or local consensus building.

Moreover, formal evaluation of the impact of guidelines on practice is rare. A systematic review of such evaluations published in the *Lancet* in 1993 identified only 42 published studies that controlled for other factors that might promote change, and involved an element of randomisation[140]. Uncontrolled before-and-after studies in individual settings are the commonest form of evaluation of the many thousands of clinical guidelines now published across the world. Thus the picture that emerges is of organisations that may invest considerable time and effort in the 'scientific' process of developing evidence-based clinical guidelines, but pay relatively little attention to the 'sociological' factors (group attitudes, financial constraints, organisational issues etc.) that govern their practical use. This is perhaps not surprising given the fact that guideline production is the easy part, while changing attitudes and systems is far more problematic. However, 'funding only the creation of guidelines orphans them' according to Dr Antonio Basinski of the Institute for Clinical Evaluative Sciences in Ontario[141].

Matters seem little different in the UK. Dr Anthony Nowlan of Hewlett-Packard, in a paper to an October 1995 conference on clinical effectiveness, notes that 'even when the motives are of the finest, the clinical value of the guidelines is not questioned, and they have been adopted in principle, there is still no guarantee of success at the point of care'[142]. Reasons include:

- **'Clinical environment**
  Pressures of time, with frequent interruptions and shifts in attention make it hard to keep a firm control over the process, encouraging slips, omissions, and mistakes. Doctors are already overloaded with information and a guideline that is not well integrated with the task can become yet another thing to attend to.
- **Scale and complexity**
  Guidelines by necessity can be large and complex, and need to be adapted on a patient by patient basis. Attempts to simplify them may result in a loss of clinical validity and produce unacceptable numbers of exceptions.
- **Localisation**
  Adjustment to locally available skills and circumstances is generally essential but this may require a significant investment of local resources.
- **Co-operation and workflow**
  Multiple clinicians and disciplines may be required to co-operate in following a guideline, making overall control problematic'.

Other barriers, identified in the Canadian survey noted above, include a lack of consensus among the medical community that guidelines are either necessary or helpful, fuelled by fears that they may compromise clinical freedom. The inclusion of cost effectiveness information may be particularly resented by doctors whose clinical focus is on the most effective treatment for the individual patient[143]. They may also resist guidelines produced by the 'wrong' experts, by groups perceived to be unsympathetic to the prevailing culture of medical practice, or by anyone without the status of local opinion leader; 'not invented here' is a major factor which needs to be taken into account, especially in the context of attempts to develop national guidelines. There may, indeed, be an inevitable built-in resistance to clinical guidelines (even those produced locally) because they are essentially remedial in intent, an 'attempt to refine decision making and to narrow practice variation to a degree unlikely to be achieved naturally by the target audience'[144]. Ensuring voluntary compliance with radical and/or remedial change is always likely to be difficult, and expecting it to happen without any positive reinforcement is flying in the face of common sense.

In order to succeed, according to Dr Nowlan, the primary purpose of clinical guidelines must be the support of clinical workers in the day-to-day job of caring for patients. They must augment and not replace clinical skills, support clinical tasks and processes, enable clinical communications, and – crucially – be 'unremarkable'. As with any good tool, clinical guidelines should come to the notice of the user only when there is a breakdown in their use, and in normal circumstances their application should be a seamless part of the working day. In other words, success depends at least in part on disguising the fact that guidelines are designed to promote significant changes in behaviour, thus keeping fear or resentment of innovation under control.

There are those who argue that the substantial investment in evidence-based clinical guidelines is never likely to produce a significant return because of the many cultural and organisational barriers to effective implementation. They are always likely to founder on the resistance of individuals to the attempts of outsiders to change their behaviour, especially in rapidly evolving areas of practice where today's expensively compiled good practice guideline may become obsolete in a relatively short time. Moreover, social scientist Jack Dowie of the Open University argues that 'practitioners, as a result of their educational and training curricula, lack the analytic skills in decision making to process systematically the evidence that they are being urged to obtain and critique'. Without these skills he believes that the promotion of EBM 'is equivalent to asking the cart to move without the horse'[145]. Educational strategies which focus on changing patterns of behaviour from within to promote continuous learning and broad-based medical decision making skills are likely to deliver a much more secure outcome, albeit more slowly.

The local dimension

However, clinical guideline development will clearly continue as an important element of the EBM movement. *Effective Health Care Bulletin* 8 on implementing clinical practice guidelines shows, on the basis of a systematic review of the evidence, that they can change medical practice and improve patient outcome[133]. Locally-based initiatives are particularly effective in ensuring commitment and effective implementation, and there may well be a case for some duplication of effort despite the efficiency savings that can be gained by the development of national guidelines. Local bodies may need to produce their own guidelines, even if these are virtually identical to the national models, to ensure ownership and increase the chances of effective dissemination and implementation within particular local contexts. However, guideline development is expensive, and local initiatives cannot hope to draw on the same level of scientific resource as those carried out at national level. It is for this reason that NHS policy is now moving towards the development of national guidelines that can be modified for local use.

There are many local/regional EBM initiatives underway such as the North of England evidence-based guidelines development project[146]. Others include:

- **ANGLIA AND OXFORD REGION**
  R&D Directorate
  Institute of Public Health
  Cambridge CB2 2SR

Tel: 01223 330167
Fax: 01223 330168
Email: mckinnelli@rdd-phru.cam.ac.uk
URL: **http://wwwlib.jr2.ox.ac.uk/a-ordd/index.htm**

The Anglia and Oxford Region lies at the centre of the EBM movement in the UK, acting as a home to the UK Cochrane Centre, the Centre for Evidence-Based Medicine and the Critical Appraisal Skills Programme. Its home page provides access to a wide range of information about the NHS R&D strategy, regional R&D developments, R&D dissemination initiatives, and EBM sources on the Internet.

Its best known EBM product is the monthly *Bandolier* newsletter, so-called because it contains 'bullet points of evidence-based medicine'. Different types of bullet – gold, silver and bronze – are recommended depending on the amount and quality of hard evidence supporting their effectiveness, and cover treatments, diagnostic tests, services and management issues. For a bullet to be fired, there must not only be research evidence that it results in the provision of a better health service, it must also be open to assessment by audit so that users can see whether it has hit the target. Bandolier is available in both priced hard copy form and on the Internet (for free, though the online version may run several months in arrears), and is now distributed in other NHS regions. It can be accessed via the URL: **http://www.jr2.ox.ac.uk:80/Bandolier/index.html**

Anglia and Oxford's best known implementation initiative is GRiPP (Getting Research into Practice and Purchasing)[147]. Initiated by the then Oxford Regional Health Authority in 1993, GRiPP involves Oxfordshire, Berkshire, Buckinghamshire and Northamptonshire and has four main aims: to identify and undertake the steps involved in using research evidence to change clinical practice; to examine the role of commissioning authorities and the contracting process in this context; to document and share lessons from projects; and to involve patients and the general public in the process.

The GRiPP initiative, which is hospital-focused, began with four key topics which were investigated separately by the participating health authorities: the use of corticosteroids in the treatment of threatened pre-term delivery (Oxfordshire), the management of stroke services (Northamptonshire), the use of D&C to treat dysfunctional uterine bleeding in women under 40 (Buckinghamshire), and the use of grommets as a treatment for glue ear (Berkshire). In the case of corticosteroids, for example, a baseline audit of current practice was undertaken by local providers. Guidelines were developed and the findings incorporated into contracts, after which a further audit of practice was conducted revealing a substantial change in practice.

According to its proponents, the GRiPP initiative offers a number of lessons for those seeking to promote an evidence-based approach to health care: projects need to address local concerns; multi-disciplinary involvement in the work is essential; guidelines must be developed by clinicians for clinicians; guidelines require both local ownership and local legitimisation; contracts cannot be a major vehicle for change, although they can formalise what has been agreed collaboratively; local opinion leaders must be used; and partnership with providers and GPs is essential.

The Anglia and Oxford R&D Directorate also supports the PRISE (Primary Care Sharing the Evidence) project which is based in the Institute of Health Sciences in Oxford, and is part of a development programme management by the Health Care Libraries Unit. It focuses on twelve primary health care sites, and aims to provide GPs and other practice-based professionals with access to good quality evidence-based information.

- **FACTS (FRAMEWORK FOR APPROPRIATE CARE THROUGHOUT SHEFFIELD)**
  Sheffield Centre for Health and Related Research
  Regent Court
  30 Regent Street
  Sheffield S1 4DA

  Tel: 0114 275 5658
  Fax: 0114 275 5653

  An initiative established in 1994 with the support of Trent Regional Health Authority to create a reproducible and cost-effective framework for changing clinical practice on the basis of research evidence[148]. FACTS has a primary care (general practice) focus, and began by exploring the management of coronary heart disease. Unlike the GRiPP initiative, it includes much less emphasis on the role of purchasers and contracts.

  Other developments in Trent can be monitored through the Regional Office home page at the URL: **http://www.netlink.co.uk/users/nhstrent/trentrd/rd.html** They include the development of a series of Health Gain Investment Programmes which are designed to guide purchasers towards making clinically and cost effective purchasing decisions. They cover a range of areas including coronary heart disease and stroke, diabetes, lung cancer and HIV/AIDS.

- **SOUTH AND WEST REGION**
  R&D Directorate
  Canynge Hall
  Whiteladies Road
  Bristol BS8 2PR

  Tel: 0117 928 7224
  Fax: 0117 928 7204)
  URL: **http://www.epi.bris.ac.uk/rd**

  The South and West Region (originally the Wessex Regional Health Authority) has established a specific structure 'to provide succinct cost-effectiveness information at a local level with a short response time'[149]. The Institute of Public Health Medicine's effectiveness team synthesises research information on both the cost and clinical effectiveness of particular health care interventions identified by both purchasers and providers as suitable for review. They are analysed according to a rigid protocol which includes topic definition; the proposed service; the current service provided to potential beneficiaries of the proposed service; numbers of patients; benefits and disbenefits; and costs and savings. Where possible, benefits and disbenefits are translated into quality adjusted life years (QALYs)[150].

Copies of commissioned reviews are available in health care libraries throughout the region, and can also be obtained from the Health Care Evaluation Unit, Department of Social Medicine, University of Bristol, Canynge Hall, Whiteladies Road, Clifton, Bristol BS8 2PR (Tel: 0117 928 7382; Fax: 0117 928 7325). The Institute of Public Health Medicine, headed by Dr Andrew Stevens, is located at Dawn House, Highcroft, Romsey Road, Winchester SO22 5DH (Tel: 01962 863511; Fax: 01962 864698; Email: andrew@wiphm.soton.ac.uk).

Each report from the effectiveness team is scrutinised by the Development and Evaluation Committee which decides if it is to be recommended to purchasers on the basis of two factors: the overall approximate cost-utility of the intervention measured in pounds per life gained or quality equivalent, and the quality of the evidence on which the effectiveness team's report is based (e.g. RCT, other controlled study, uncontrolled studies, current opinion). In general, moderate cost-utility combined with firm evidence, or high cost utility with weaker evidence, would produce a 'strongly recommended' decision.

Recommendations are forwarded to purchasing authorities. Those that are 'strongly recommended' are expected to be implemented via regional contracts; 'not recommended' interventions should be dropped from the purchasing agenda; while 'recommended' and 'beneficial but high cost' interventions are left to local discretion. Some Development and Evaluation Committee reports are available on the Internet (**http://www.epi.bris.ac.uk/rd/publicat/dec/index.htm**), while all can be obtained in hard copy from the R&D Directorate, South and West Regional Health Authority, Canynge Hall, Whiteladies Road, Bristol BS8 2PR (Tel: 0117 928 7224; Fax: 0117 928 7204).

The South and West Region also distributes the Anglia and Oxford Region's bulletin, *Bandolier*, to health care professionals in the South West, and has a publishing programme of its own. This includes *Briefing Papers* offering 'authoritative discussions on the effectiveness of health interventions' (**http://www.epi.bris.ac.uk/rd/publicat/briefing/index.htm**) and a quarterly *Newsletter* which covers all aspects of its R&D activity. Between March 1994 and November 1996, it also published a bi-monthly digest of new research of interest to health care commissioners and clinicians interested in EBM. Called *Evidence-Based Purchasing*, it covered a wide range of health care subjects as well as more general EBM issues such as the role of clinical practice guidelines, the work of the Cochrane Collaboration and the NHS Centre for Reviews and Dissemination, and social science and evidence-based purchasing.

- **WEST MIDLANDS REGION**
  Aggressive Research Intelligence Facility
  27 Highfield Road
  Edgbaston
  Birmingham B15 3DP

  Tel/Fax: 0121 455 6852
  URL: **http://www.hsrc.org.uk/links/arif/arifhome.htm**

  ARIF is a collaborative effort set up by the West Midlands Region's R&D Department and involving the University of Birmingham's Department of Public Health and

Epidemiology, Department of General Practice and Health Services Management Centre (the last of these has recently published a report on the development of clinical guidelines in the West Midlands[151]). It consists of a team of three people whose aim is 'advancing the use of evidence on the effects of health care in the West Midlands'. Funded for three years from 1 July 1995, ARIF's first objective is providing 'timely access to, and advice on, existing reviews of research', primarily for purchasing authorities which are allowed to submit one request per calendar month. A pilot service to a small number of GP fundholders is also underway, and ARIF will respond sympathetically to requests for help from other quarters as resources permit.

Although only a small scale initiative as yet, ARIF has plans for expansion in view of the level of demand for its services. It also hopes to develop much closer links with the NHS Centre for Reviews and Dissemination (in particular by contributing to the DARE database); maintain and expand its library of systematic reviews; develop its service to purchasing organisations; develop a programme of workshops on searching and critical appraisal skills; and develop a research agenda focusing on primary research into the effectiveness of interventions to improve the application of research evidence (particularly the role of research dissemination units such as ARIF itself).

Other developments in the West Midlands Region can be accessed via its home page at the URL: **http://www.gold.net/users/ei26/index.htm**

- **PACE (PROMOTING ACTION ON CLINICAL EFFECTIVENESS)**
  King's Fund
  11-13 Cavendish Square
  London W1M 0AN

  Tel: 0171 307 2653

This new initiative from the King's Fund also has a specifically local dimension. It was set up in late 1996 as a pathfinder programme for the successful implementation of evidence-based practice in the NHS and will initially involve 16 projects around the UK which are designed to use research-based information to improve services to patients. These are being selected from applications submitted by health authorities and trusts, and will run for two years.

In addition, the King's Fund hopes to create a PACE Network to link people interested in the promotion of evidence-based practice. This will be open to all NHS disciplines and functions including practitioners, managers, researchers and health economists. The first year's membership is free, and the King's Fund plans a number of discussion days, focusing on different EBM issues, to share ideas, experiences and problems.

(120)
CLINICAL audit enabler: a clinical resource reference guide
Gailey, F (editor)
Lothian Clinical Audit: Edinburgh Mar 1996. 122pp
There is a huge literature on clinical audit, both general and in relation to individual health care activities. No attempt is made here to cover it, but this directory gives a useful introduction to some of the many organisations involved, not only in Scotland but

elsewhere in the country. It includes sections on national clinical audit centres; institutional sources for clinical outcomes studies and clinical guidelines; the Royal Colleges and national professional associations in medicine, dentistry, nursing and professions allied to medicine; research organisations (including CASPE Research, the NHS Centre for Reviews and Dissemination and the UK Cochrane Centre); and educational organisations. Also includes descriptions of information and library resources within the Lothian area, examples of resources available from pharmaceutical companies, and a summary of clinical audit and related work by Lothian Clinical Audit.

(121)

MEDICAL audit activities of the Royal Colleges and their faculties in the UK
Hopkins, A (editor)
Department of Health, Richmond House, 79 Whitehall, London SW1A 2NS, 1994. 142pp
For a further evaluation of Royal College activities
See also: Evaluating audit: the audit activities of the Medical Royal Colleges and their faculties in England, by M Amess and others (Clinical Accountability Service Planning and Evaluation (CASPE) Research, 76 Borough High Street, London SE1 1LL, Apr 1995. 141pp)

(122)

THE DEVELOPMENT of audit: findings of a national survey of healthcare provider units in England
Buttery, Y and others
Clinical Accountability Service Planning and Evaluation (CASPE) Research, 76 Borough High Street, London SE1 1LL, 1994. 153pp
For further CASPE reports
See also: Evaluating audit: developing a framework, by K Walshe and J Coles (CASPE Research, Jul 1993. 66pp)
Evaluating audit: a review of initiatives, by K Walshe and J Coles (CASPE Research, Jul 1993. 61pp)
The role of the commissioner in audit: findings of a national survey of commissioning authorities in England, by M Rumsey and others (CASPE Research, 1994. 73pp)
Provider audit in England: a review of 29 programmes, by Y Buttery and others (CASPE Research, 1995. 182pp)
A review of audit activity in the nursing and therapy professions: findings of a national survey, by M Wilmot and others (CASPE Research, 1995. 100pp)

(123)

CLINICAL audit: meeting and improving standards in health care
Department of Health, Richmond House, 79 Whitehall, London SW1A 2NS, 1993. 14pp
For further Department of Health reports on audit
See also: Medical audit: working paper 6 (Department of Health: London, 1989. 13pp)
Medical audit and information technology: report of a working group of the Clinical Resources and Audit Group (Department of Health, 1990. 45pp)
Clinical audit in the nursing and therapy professions, by K Von Degenberg (Department of Health, 1993. 185pp)
The evolution of clinical audit (Department of Health, 1994. 29pp)
Medical audit in the hospital and community health service, by M Stern and S Brennan (Department of Health, 1994. 189pp)

Clinical audit in primary health care (Department of Health, Jul 1994. 62pp)
Medical audit in primary care: a collation of evaluative projects 1991-1993: a report for the
NHS Management Executive of the Department of Health, by C Humphrey and D
Berrow (Department of Health, 1995. 63pp)

(124)

**THE NHS and the clinical audit initiative: outline of planned monitoring
arrangements**
NHS Management Executive, Richmond House, 79 Whitehall, London SW1A 2NS,
1995. (EL(95) 103)

(125)

CLINICAL audit in England
National Audit Office
HMSO (now Stationery Office), Dec 1995. 51pp (HC 1995-96 27)
For the companion report from the House of Commons Public Acounts Committee, and
for the 1994 reports on clinical audit in Scotland
See also: National Health Service Executive clinical audit in England: 31st report, session
1995-96, by House of Commons Committee of Public Accounts (HMSO, Jul 1996. 41pp.
HC 1995-96 304)
Auditing clinical care in Scotland, by National Audit Office (HMSO, 1994. 23pp. HC
1993-94 275)
Auditing clinical care in Scotland: 51st report, session 1993-94, by House of Commons
Public Accounts Committee (HMSO, 1994. 29pp. HC 1993-94 375)

(126)

**SPECIALISED services (1996): an overview and summary of CSAG's studies
since 1992**
West, E and others
Clinical Standards Advisory Group, NHS Executive, HCD-PH1B Room 409,
Wellington House, Waterloo Road, London SE1 8UG, Aug 1996. 72pp
Covers services dealing with cystic fibrosis, childhood leukaemia, neonatal intensive care
and coronary revascularisation.

(127)

REPORT on confidential enquiries into maternal deaths in the UK 1988-1990
Department of Health and others
HMSO (now Stationery Office), 1994. 198pp
An example of one of the specialised national clinical audits carried out in recent years.
Others cover perioperative deaths; stillbirths and deaths in infancy; genetic services; major
trauma outcomes; intensive care; the Clinical Oncology Information Network; the
National Cardiology/Cardiothoracic Surgery Database; and hospital infection control.
Contact details are provided in Annex 3b of *Promoting clinical effectiveness*[62, Ch.2].

(128)

AUDIT of Medical Audit Advisory Groups
Humphrey, C; Berrow, D
Department of Health, Richmond House, 79 Whitehall, London SW1A 2NS, 1993.
For a national survey of MAAGs

See also: National MAAG survey 1992-93, by Birmingham Medical Audit Advisory Group (Department of Health, 1994.)

(129)
SOME reservations about clinical guidelines
Hopkins, A
Archives of Disease in Childhood, Jan 1995 72(1) pp70-75, 34 references
Looks at the origins of the 'current fashion' for clinical guidelines, and reviews a wide range of problems associated with them including the tendency to write guidelines for clinical diagnosis rather than for a presenting problem; the failure to identify target health professional audiences clearly; the failure to identify target patient populations clearly; the bias of guidelines towards technical aspects of care as opposed to interpersonal and other aspects which are equally important, especially in the context of chronic disease; the failure to reflect user perspectives; the limitation of most guidelines to single disorders when many patients, especially older people, have more than one active illness; and the mistaken belief that guidelines written for health professionals can be used to determine the contracting process. For earlier American comment on the deficiencies of many clinical guidelines
See also: Medical practice guidelines: current activities and future directions, by A M Audet (Annals of Internal Medicine, 1 Nov 1990 113(9) pp709-14)

(130)
MAKING decisions about health technologies: a cost-effectiveness perspective
Rutten, F; Drummond, M
University of York Centre for Health Economics, Hesington, York YO1 5DD, Jan 1994. 95pp
A handbook to inform policy makers about the techniques of economic appraisal, and help them in comparing the benefits and opportunity costs of alternative strategies for the allocation of resources. Discusses the role of economic appraisal in general terms, and includes case studies relating to national breast screening programmes, the evaluation of drugs for the management of labour, and the planning of heart transplantation services. For more on the economic evaluation of health care, including a more recent analysis by Drummond and colleagues
See also: Increasing the impact of economic evaluations on health-care decision-making, by D Coyle (University of York Centre for Health Economics, Jul 1993. 39pp. Discussion Paper 108)
Economic evaluation and health care: cost-utility studies, by R Robinson (British Medical Journal, 2 Oct 1993 307(6908) pp859-62, 26 references)
Economic evaluation and health care: the policy context, by R Robinson (British Medical Journal, 16 Oct 1993 307(6910) pp994-96, 14 references)
Economic evaluation in health care decision making: evidence from the UK, by M Drummond and others (University of York Centre for Health Economics, Apr 1996. 38pp. Discussion Paper 148)

(131)
REGISTER of cost effectiveness studies
Department of Health, Richmond House, 79 Whitehall, London SW1A 2NS, Aug 1994. 122pp
For an evaluation of the Register, and for the NHS Centre for Reviews and Dissemination guidance on input to the NHS Economic Evaluation Database
See also: The Department of Health Register of Cost Effectiveness Studies: a review of study content and quality, by J Mason and M Drummond (University of York Centre for

Health Economics, Hesington, York YO1 5DD, and University of York Health Economics Consortium, Feb 1995. 33pp. Discussion Paper 128)
Making cost-effectiveness information accessible: the NHS Economic Evaluation Database project: guidance for reporting critical summaries of economic evaluations, by A Vanoli and T A Sheldon (NHS Centre for Reviews and Dissemination, University of York, Sep 1996. 71pp. CRD Report 6)

(132)
GUIDELINES for clinical practice: from development to use
Field, M J; Lohr, K N
National Academic Press: Washington, DC, 1992. 426pp
Identifies nine desirable attributes of clinical guidelines which should, where possible, include a cost effectiveness element: validity, reproducibility, reliability, representative development, clinical applicability, clinical flexibility, clarity, meticulous documentation, and scheduled review.

(133)
IMPLEMENTING clinical practice guidelines: can guidelines be used to improve clinical practice?
Grimshaw, J M; Russell, I T
University of Leeds Nuffield Institute of Health, and others, 1995. 11pp (Effective Health Care Bulletin 8)
Produced in association with the University of Leeds Centre for Health Economics, the NHS Centre for Reviews and Dissemination, and the Royal College of Physicians Research Unit. Argues on the basis of a systematic review of the evidence that clinical guidelines can have a positive impact on practice.

(134)
HOW should information on cost effectiveness influence clinical practice?
Williams, A
In: Outcomes into clinical practice, edited by T Delamothe (BMJ Publishing: London, 1994 pp99–107)
Argues that information on cost effectiveness should influence clinical practice because resources for health care are limited. This is accepted in principle by the NHS and other official bodies but apparently forgotten in practice, perhaps because of its potential sensitivity both for patients and health care professionals. The pursuit of clinical excellence whatever the cost is an unrealistic objective, and it is essential that cost effectiveness information becomes an integral part of clinical practice guidelines.

(135)
ECONOMIC evaluation in health care research and development: undertake it early and often
Sculpher, M and others
Brunel University Health Economics Research Group, Mar 1995. 28pp, bibliography + list of HERG publications (Discussion Paper 12)
Argues that economic evaluations of health care interventions are developing a key role in health care R&D but that their full scope has yet to be realised. In particular, there is little appreciation by public research funders and pharmaceutical companies of the appropriate design and timing of economic evaluation, and little understanding of its potential strategic role in setting research priorities and maximising returns from R&D investment.

Recommends the early and iterative use of economic evaluation as an integral part of the R&D process.

(136)

CLINICAL guidelines: using clinical guidelines to improve patient care within the NHS

NHS Executive, Richmond House, 79 Whitehall, London SW1A 2NS, Jul 1996. 30pp
Summarises the NHS Executive's approach to the development and appraisal of national clinical guidelines for use in the NHS, including comment on legal considerations, and the appraisal, dissemination, implementing and monitoring of guidelines. The approach has been developed in the context of *Effective Health Care Bulletin* 8 and the guidance of the Clinical Outcomes Group, a multi-professional group that advises the Department of Health on how to improve the outcomes of clinical care. Support has also been secured from the Clinical Standards Advisory Group, the Standing Medical Advisory Committee, the Standing Nursing and Midwifery Advisory Committee and the Joint Consultants Committee. For an earlier report from the Clinical Resource and Audit Group in Scotland
See also: Clinical guidelines, by Clinical Resource and Audit Group (Scottish Office: Edinburgh, May 1993. 78pp)

(137)

THE DEVELOPMENT and implementation of clinical guidelines: report of the Clinical Guidelines Working Group

Royal College of General Practitioners, 14 Princes Gate, Hyde Park, London SW7 1PU, Apr 1995. 31pp (Report from General Practice 26)
This is included as an example of Royal College clinical guidelines initiatives. The RCGP, like other professional bodies, is concerned to ensure the development of sound clinical practice guidelines, and identify strategies for their successful implementation. Its report deals with the pitfalls and obstacles to guideline effectiveness; guideline development; ensuring that guidelines change clinical practice; and the implementation of guidelines in general practice. Appendices cover the commissioning strategy document and a draft appraisal instrument for clinical guidelines. For another example of a Royal College initiative
See also: Clinical practice guidelines and their development (Royal College of Psychiatrists, 17 Belgrave Square, London SW1X 8PG, Jun 1994. 45pp. Council Report CR34)

(138)

MAKING clinical policy explicit: legislative policy making and lessons for developing practice guidelines

Lomas, J
International Journal of Technology Assessment and Health Care, Winter 1993 9(1) pp11-25
Argues that clinical policy making is analogous to legislative policy making in that both are concerned with rules governing the behaviour of individuals or institutions. Clinical practice guidelines emphasise the value of making clinical policy making more explicit, and it is argued that experience from the legislative field could help in the guideline development process. Develops generic steps for making decisions with incomplete information; synthesising facts, vested interests and values; involving stakeholders; and implementing policy. For more from Jonathan Lomas
See also: A taxonomy and critical review of tested strategies for the application of clinical practice recommendations: from official to individual clinical policy, by J Lomas and R B

Haynes (American Journal of Preventive Medicine, 1988 4(4 supplement) pp77-94 + discussion pp95-97)

Do guidelines guide practice? Effect of a consensus statement on the practice of physicians, by J Lomas and others (New England Journal of Medicine, 9 Nov 1989 321(19) pp1306-11)

Opinion leaders vs. audit and feedback to implement practice guidelines, by J Lomas and others (British Medical Journal, 17 Aug 1991 303(6799) 398-402)

(139)
REPORT on activities and attitudes of organizations active in the clinical practice guidelines field

Carter, A O and others

Canadian Medical Association Journal, 1 Oct 1995 153(7) pp901-07, 7 references

The first in a series of six articles reporting on the 1994 Canadian Clinical Practice Guidelines Network Workshop. Presents the results of a survey of invited organisations on their opinions about the five workshop themes: organisational roles in clinical guideline development; priority setting; guideline implementation; guideline evaluation; and the role of the CPG Network. For Anne Carter's summary review of the Workshop proceedings

See also: Proceedings of the 1994 Canadian Clinical Practice Guidelines Network Workshop, by A O Carter and others (Canadian Medical Association Journal, 15 Dec 1995 153(12) pp1715-19, 15 references)

Note: as a key part of its initiative on clinical practice guidelines, the Canadian Medical Association is developing the CPG Infobase, a database of clinical guidelines that have been endorsed by national, provincial or territorial medical or health organisations, professional societies, government agencies or expert panels. Those already published in the Canadian Medical Association Journal are available in full text, while others are listed and will be added once the developers have granted permission. Additional information is available from Sue Beardall, Senior Project Manager, Quality of Care Program, Health Care and Promotion, Canadian Medical Association, PO Box 8650, Ottawa ON K1G 0G8. The CPG Infobase can be accessed via **http://www.cma.ca/cpgs/index.html**. The Canadian government has also set up a National Forum on Health (PO Box 2798, 200 Kent Street, 4th Floor, Ottawa, Ontario, Canada K1P 6H4) which includes evidence-based decision making as one of its four major themes. It can be accessed via **http://www.nfh.hwc.ca/index.html**.

(140)
EFFECT of clinical guidelines on medical practice: a systematic review of rigorous evaluations

Grimshaw, J M and others

Lancet, 27 Nov 1993 342(8883) pp1317-22

Reviews 59 published evaluations of clinical guidelines that met defined criteria for scientific rigour: 24 dealt with guidelines for specific clinical conditions, 27 with preventive care, and 8 with prescribing or support services. Fifty-five studies reported significant improvement in the process of care after the introduction of guidelines, while 9 of the 11 studies which assessed the outcome of care also reported improvements. Concludes that explicit guidelines do improve clinical practice when introduced in the context of rigorous evaluation, but that the size of improvements varies considerably. For further comment from Jeremy Grimshaw and colleagues

See also: Achieving health gain through clinical guidelines 1: developing scientifically validated guidelines, by J M Grimshaw and others (Quality in Health Care, Dec 1993 2(4) pp243-48)

Achieving health gain through clinical guidelines 2: ensuring guidelines change medical practice, by J M Grimshaw and others (Quality in Health Care, Mar 1994 3(1) pp45-52)

Critical appraisal of clinical practice guidelines: a pilot study of an evaluation instrument, by F Cluzeau and others (St George's Hospital Medical School Health Care Evaluation Unit: London, 1994. 107pp)

Appraising clinical guidelines: towards a 'Which' guide for purchasers, by R S Hayward and others (Quality in Health Care, Sep 1994 3(3) pp121-22)

Clinical practice guidelines: do they enhance value for money in health care? by J M Grimshaw and A Hutchinson (British Medical Bulletin, 1995 51(4) pp927-40, 33 references)

(141)
EVALUATION of clinical practice guidelines
Basinski, A S H
Canadian Medical Association Journal, 1 Dec 1995 153(11) pp1575-81, 40 references
Relatively little attention has been devoted to the evaluation of clinical guidelines despite the many thousands that have now been produced. Identifies three types of evaluation that might be used: evaluation of guidelines under development and before dissemination and implementation; evaluation of health care programmes in which guidelines play a significant role; and scientific evaluation through studies providing the scientific basis for the further development of guidelines.
Note: Dr Basinski is senior scientist at the Institute for Clinical Evaluative Sciences, Room G-2, 2075 Bayview Avenue, North York, Ontario M4N 3M5. This is a non-profit organisation dedicated to research that contributes to the effectiveness, quality and efficiency of health care in Ontario. It produces a newsletter available via **http://www.ices.on.ca/**

(142)
INFORMATION systems and the support of good practice
Nowlan, A
In: Clinical effectiveness: from guidelines to cost-effective practice, edited by M Deighan and S Hitch (Earlybrave Publications Limited, PO Box 3165, Brentwood, Essex CM13 1TL, Oct 1995 pp23-28)
Discusses the many barriers that exist to the implementation of clinical guidelines, and looks at how they might be overcome. For other views on how best to get evidence into practice
See also: How to ensure that guidelines are effective, by R Thomson and others (British Medical Journal, 22 Jul 1995 311(6099) pp237-42)
Professional roles in promoting evidence-based practice, by G Batstone and M Edwards (British Journal of Health Care Management, Mar 1996 2(3) pp144-47, 11 references)

(143)
EVIDENCE-BASED medicine meets cost-effectiveness analysis
Clancy, C M; Kamerow, D B
Journal of the American Medical Association, 24/31 Jul 1996 276(4) pp329-30, 11 references

Looks at the possible contradictions between clinical and cost effectiveness information, noting that each is produced for a different audience (clinicians and policymakers) with different aims and objectives.

(144)
PARADOX, process and perception: the role of organizations in clinical practice guidelines development
Lewis, S
Canadian Medical Association Journal, 15 Oct 1995 153(8) pp1073-77, 14 references
Despite the importance of organisations in the development of clinical guidelines, their role has been virtually ignored by analysts who have tended to see guidelines as strictly rational, scientific entitities. The critically important sociological factors affecting guideline acceptance and implementation are discussed, with comment on the paradoxes raised by a clash between the scientifically modelled guideline and the values of medical practice.

(145)
'EVIDENCE-BASED', 'cost effective' and 'preference-driven' medicine: decision analysis based medical decision making is the pre-requisite
Dowie, J
Journal of Health Services Research and Policy, Apr 1996 1(2) pp104-13, 40 references
A social scientist's view of the EBM movement which argues that decision analysis based medical decision making (DABMDM) is an essential prerequisite of the move to an effectiveness-based NHS. DABMDM allows the clinician to incorporate all the key elements of good clinical decision making – not just the best scientific evidence (which dominates EBM) but cost-effectiveness information and the values and preferences of patients.

(146)
DEVELOPING valid guidelines: methodological and procedural issues from the North of England Evidence Based Guidelines Development Project
Eccles, M and others
Quality in Health Care, Mar 1996 5(1) pp44-50
For further comment
See also: North of England evidence based guidelines development project: methods of guideline development, by M Eccles and others (British Medical Journal, 23 Mar 1996 312 (7033) pp760-62)

(147)
GETTING Research into Practice and Purchasing (GRiPP): four counties approach: resource pack
Anglia and Oxford Regional Health Authority: Oxford, and NHS Executive, Jun 1994
124pp
For more on the GRiPP initiative, initially called GRiP (Getting Research into Practice)
See also: Getting research into practice: facing the issues, by S Dopson and others (Journal of Management in Medicine, 1994 8(6) pp4-12)
Getting a GRiP, by M Dunning, M (Health Service Journal, 28 Apr 1994 104(5400) pp24-26)
A GRiPPing yarn: getting research into practice: a case study, by G Needham (Health Libraries Review, Dec 1994 11(4) pp269-77, 7 references)
Bringing it together: the GRiPP experience, by J Meara and N Hicks (In: Clinical effectiveness: from guidelines to cost-effective practice, edited by M Deighan and S Hitch.

Earlybrave Publications Limited, PO Box 3165, Brentwood, Essex CM13 1TL, Oct 1995 pp123-26)

(148)
FACING the FACTS
Munro, J
Health Service Journal, 5 Oct 1995 105(5473) pp26-27

(149)
COST-EFFECTIVENESS and health care purchasing
Stevens, A
In: Clinical effectiveness: from guidelines to cost-effective practice, edited by M Deighan and S Hitch (Earlybrave Publications Limited, PO Box 3165, Brentwood, Essex CM13 1TL, Oct 1995 pp119-21, 9 references)
Briefly describes the activities of the South West Region in providing cost-effectiveness information at a local level.

(150)
HEALTH technology evaluation: research reviews
Stevens, A (editor)
Wessex Institute of Public Health Medicine, Dawn House, Highcroft, Romsey Road, Winchester SO22 5DH, and NHS Executive, 1994. 2 volumes
Further reviews now available in hard copy as outlined above.

(151)
IMPROVING clinical effectiveness: the development of clinical guidelines in the West Midlands
Honigsbaum, F; Ham, C
University of Birmingham Health Services Management Centre, 40 Edgbaston Park Road, Birmingham B15 2RT, 1996. 56pp
Apart from its work on guidelines, the Health Services Management Centre is also heavily involved in research into ways of involving the public and patients in health care planning, prioritising and rationing. In addition, its current work programme includes the development of a database of reports and other materials on world wide approaches to priority-setting in health care. Details of all HSMC's publications are available on application, and it also publishes a regular newsletter, now available on its web site at **http://www.bham.ac.uk/hsmc/**

4.4 The role of libraries

Evidence-based medicine places information searching, evaluation and synthesis at the heart of the health service, highlighting both traditional and computer-based skills in searching for RCTs and other reputable scientific evidence. These skills lie – or should lie – at the heart of professional librarianship whose primary function is to help those who need information to extract relevant material from the huge mass of documentation (in both conventional and electronic forms) that faces them. The potential importance of specialist health care librarians in promoting and assisting the development of EBM was recognised in the two 'Cumberlege seminars' of 1992 and 1993 which prompted the Department of Health to establish the post of NHS Library Adviser to implement some of the seminar proposals[152,153].

The seminars focused on the information implications of the massive changes introduced into the NHS in the previous ten years including the downsizing and reorganisation of the NHS Executive; the shift towards a primary care-led NHS and the decentralisation of decision making; the introduction of problem-based, self-directed learning in undergraduate medical training; and the growing trend towards patient involvement in health care decision making and evidence-based practice. These changes have created a variety of new information needs, summarised by former NHS Library Adviser Margaret Haines as follows[154].

- 'mechanisms for efficient transfer of information between NHS staff and organisations
- sources of information about clinical effectiveness at point of care
- information resources suitable for patients and their families
- quality filters to assist rapid retrieval from Internet and other networks
- better indexing and coverage of grey literature in existing databases
- understanding of the needs of new NHS staff such as GP fundholders'

However, despite the clear relevance of library and information skills to meeting these needs, Margaret Haines' introduction to a special issue of *Health Libraries Review* in December 1994 laments the fact that 'most librarians, if they are interested, do not seem to be visible or proactive in the process'[155]. The library profession is, unfortunately, rather prone to undervaluing (or failing to recognise) its potential contribution to the information-intensive, learning organisation but, unless it does, it will 'lose out on a golden opportunity to demonstrate just how important and relevant our particular information handling skills are to the whole process of improving the quality of health care'. The Cumberlege seminars identified several reasons for the failure of NHS librarians to maximise their potential including inadequate and complex funding regimes, lack of integration into NHS IT networks, and lack of training in the new skills required to support evidence-based health care.

However, this rather gloomy picture is lightened by the knowledge that librarians are at the forefront of EBM development in some areas. Several of the initiatives already described include participation by librarians and information professionals including the NHS Centre for Reviews and Dissemination and the UK Cochrane Centre where they are involved in the support of systematic review and the provision of enquiry answering services. Chapter 3 notes the current awareness/gatekeeper services provided by Andrew Booth at the Sheffield Centre for Health and Related Research (SCHARR) and Chris Cox at the University of Hertfordshire Library. Both maintain a watching brief on EBM Internet developments, while the OMNI service headed by the Library at the National Institute for Medical Research is the UK's most comprehensive gateway to health information on the Internet.

The team developing the ARIF (Aggressive Research Intelligence Facility) in the West Midlands (see Section 4.2 above) includes an information manager with experience in public and health libraries, while the AuRACLE (Automated Retrieval Assistant for CLinically-relevant Evidence) project in Sheffield is a joint venture between SCHARR and the University of Sheffield Department of Information Studies (see Section 4.1 above). The GRiPP initiative (see Section 4.3 above) also includes librarian involvement, as does CASP and the Centre for Evidence Based Medicine's training workshops (see Section 4.2 above). These last three initiatives are all based in the Oxford region where the Health Care Libraries Unit has played a major role in developing ways for librarians in the four associated counties (Oxfordshire, Northamptonshire, Berkshire and Buckinghamshire) to support health care professionals[156].

During Margaret Haines' two years work as NHS Library Adviser, she noted many encouraging developments. Internet and NHS Net links have been installed in many libraries to give better access to information sources, and NHS Regional Library Adviser posts have been created to improve coordination of library provision at local level (names and contact details are given in Annex 3 of *Clinical guidelines* published by the NHS Executive[136]). The NHS R&D Directorate has also funded research on primary care and consumer health information needs as well as promoting the role of libraries in clinical effectiveness. Health care librarians themselves have also been active in producing new guides to EBM sources, both print and electronic, setting up web sites on regional library networks, and developing accreditation systems for libraries to improve standards of service.

Training is a particularly important issue. Librarians working in support of evidence-based practice need to become much more discriminating about the information they supply or help their users to find, a lesson which might usefully be learned by all librarians in an information role given the huge volume of published information of widely differing quality that now exists. 'Quick and dirty' literature searches (online or manual) can do more harm than good in the EBM context, and critical appraisal training may now be essential for health care librarians. The innovative Librarian of the 21st Century project funded by the Anglia and Oxford region has encouraged several regions to provide training for librarians in these new skills.

Librarians and information workers in the health service can (and do) contribute to EBM in a variety of ways:

- At the data collection stage of the systematic review process. Librarians with sophisticated computer searching skills are well placed (in cooperation with systematic reviewers who have detailed subject knowledge) to overcome the indexing deficiencies of databases such as MEDLINE and EMBASE, and to contribute to their improvement. They can also provide guidance on the best methods of conducting manual searches of journals, and on the best sources for non-journal material including conference reports, 'grey' (report) literature and research in progress. Information staff experienced in particular medical fields can also assist in the preliminary weeding of material based on clear selection criteria. Many systematic reviews produce large volumes of material that needs to be organised systematically and, in some cases, indexed. Librarians can provide advice on the design of bibliographic databases and on indexing to ensure efficient retrieval and manipulation of material.
- The librarian's searching skills and knowledge of information sources can also be used in support of the individual practitioner wishing to conduct a critical appraisal of a particular topic or identify systematic reviews already carried out. Support may be direct (i.e. a full enquiry answering service including MEDLINE and other searches) or indirect (including advice on search strategies and potentially useful sources). Librarians may also be able to counter the 'narrow scientism' of the current EBM movement, if

this is perceived to be a problem, by identifying and offering access to complementary sources of evidence that are softer and more qualitative. Associated services could include basic and advanced training courses in database and Internet searching skills, and simple leaflets and bibliographies giving access details for the major EBM sources that the individual practitioner might use.

- Librarians can also perform an invaluable current awareness role through constant monitoring of information sources (including the Internet), and by identifying and passing on the really crucial information to those who need it. Active dissemination strategies can cover both new EBM products (systematic reviews, for example) and information sources, with current awareness bulletins or other services carefully targeted to particular audiences including researchers, practitioners, managers and the public. Presenting information in attractive, concise formats for different audiences is just as important as identifying and evaluating it. For those librarians working with researchers, policy makers, managers or clinicians, the key issue will be to challenge information overload problems by designing products in such a way that they become essential information resources rather than something which is consigned to the files or the wastepaper bin. For those working in consumer health information services, the aim will be to present sometimes difficult technical information in such a way that it is meaningful to the lay person. The issues involved in sharing information with consumers have been addressed by Bob Gann of the Help for Health Trust and Gill Needham of Buckinghamshire Health Authority in a one day course designed specifically for health care librarians[157].

- Identifying and disseminating potentially useful information are key elements in the EBM process, but the library's role does not stop there. Despite the attention paid to sophisticated computer-based information retrieval systems in EBM, many people will both need and want access to material in hard copy. The maintenance of a traditional well organised and easily accessed collection of books, journals, reports and other material – combined with efficient links to other sources both within the NHS and outside it – remains crucial. The Regional Library Adviser posts will do much to reduce unnecessary duplication of stock and maximise the use of material through cooperation.

Despite the progress made in the last two or three years, many problems still remain including inadequate funding, the lack of multidisciplinary library and information services in many institutions, and the total lack of provision for many people working in primary and community care. These issues are now being addressed by a Cross-Directorate working party within the NHS, and there is evidence to suggest that senior NHS management is becoming more aware of the value of libraries in supporting clinical- and cost-effectiveness. Continued progress will depend on librarians demonstrating in practical ways the case for treating their functions and skills as a key strategic resource. This will require continuous learning as well as a much more proactive approach than many may be accustomed to, in the context both of service to customers and relationships with the rest of the organisation. Given the traditionally low status accorded to the library/information function, this may not be easy but it must be attempted. Otherwise the key role of providing access to the information that underlies a knowledge-based health service will be carried out less effectively and efficiently by others with more limited skills.

(152)
REPORTS of a seminar on health care information in the UK, 1 July 1992 at the King's Fund Centre London, chaired by Baroness Cumberlege
The British Library, Research and Development Department: London, 1992. 41pp
(British Library R&D Report 6089)

(153)

REPORT of a seminar on managing the knowledge base of health care, 22 October 1993 at the King's Fund Centre London, chaired by Baroness Cumberlege

The British Library Research and Development Department: London, 1994. 69pp (British Library R&D Report 6133)

For a follow-up survey by the British Library, and for brief reports on Cumberlege follow-up seminars in provincial centres

See also: Managing the knowledge base of healthcare: follow-up survey, by A Beevers (British Library Research and Development Department, 1995. 36pp. Report 6182)

Managing the knowledge base of health care: the Scottish dimension, 15 Jun 1994, Scottish Health Service Centre, Edinburgh, by A Jamieson (Health Libraries Review, Dec 1994 11(4) pp294-96)

Cumberlege in the North: managing the knowledge base of health care, 29 June 1994, Civic Centre, Newcastle-upon-Tyne, by S Childs (Health Libraries Review, Dec 1994 11(4) pp296-97)

Managing the knowledge base of health care: a Welsh perspective, 25 July 1994, University of Wales College of Medicine, Cardiff, by J Lancaster (Health Libraries Review, Dec 1994 11(4) pp297-98)

(154)

THE NHS Library Adviser in England

Haines, M

IFMH Inform, 1995 6(1) pp4-6

Looks briefly at the challenges facing the NHS library and information community, with comment on the recommendations of the Cumberlege seminars and the role of the NHS Library Adviser in implementing them. For a detailed analysis published in the same year of the value of NHS library and information services to clinical decision making

See also: The value to clinical decision making of information supplied by NHS library and information services, by C J Urquhart and J B Hepworth (British Library Research and Development Department: London, 1995. 276pp. Report 6205)

(155)

EVIDENCE-BASED practice: new opportunities for librarians

Haines, M

Health Libraries Review, Dec 1994 11(4) pp221-25, 6 references

An editorial introduction to a special issue of the journal on the role of librarians in EBM. Articles (all of which are included elsewhere in this review) cover the NHS information strategy; the role of the Cochrane Centre in systematic review (with particular reference to the inadequacies of database indexing); the role of the NHS Centre for Reviews and Dissemination; the role of the UK Clearing House on the Assessment of Health Outcomes; the need for a patient-focused approach to outcomes research; the King's Fund Centre programme of research on consumer health information services; the role of librarians in the GRiPP project and other initiatives in the Anglia and Oxford Region.

(156)

FINDING the evidence

Palmer, J

Health Libraries Review, Dec 1994 11(4) pp282-86, 18 references

The Director of the Health Care Libraries Unit based at the John Radcliffe Hospital in

Oxford describes the involvement of librarians in EBM developments in the Oxford region. For additional comment on the Librarian of the 21st Century project
See also: Good diagnosis for the twenty-first century, by J Palmer and others (Library Association Record, Mar 1995 97(3) pp153-54)

(157)
SHARING outcomes information with consumers: a new course for health librarians
Entwistle, V and others
Health Libraries Review, Dec 1994 11(4) pp279-82, 7 references
Brief report on a course developed by Bob Gann and Gill Needham, and first presented to participants at the fourth European Conference for Medical and Health Libraries held in Oslo in June 1994. After an introductory session, participants split into two groups to discuss the kinds of information needed for clinical decision making from both the doctor and patient viewpoint, using a specific case study.

5. THE PATIENT PERSPECTIVE

'Patients', according to the NHS Management Executive in *Promoting clinical effectiveness*[62, Ch.2], 'can be powerful agents for change within the health service [and] sustained improvements in clinical effectiveness can only be secured by a partnership involving clinicians, patients and managers'. The 1989 White Paper *Caring for people* set out to enable people to 'achieve maximum independence and control over their own lives'[158], while the *Priorities and planning guidance* for 1996-97 requires purchasers to demonstrate 'how they use information to help patients have a better understanding of the effectiveness of various treatments'[63, Ch.2]. There is increasing acceptance of the importance of giving patients, their families and the wider public 'clear information in simple language, free of jargon and technicality'. The need to involve patients and the public in the future development of a resource-constrained NHS is also widely recognised by the medical profession, consumer bodies, patients' representative groups and health policy research bodies.

A variety of factors has coincided to produce a greater emphasis on informing and 'empowering' patients. They include a general increase in the public awareness of science as a result of improving educational levels and the activities of the media which have not only broadened the dissemination of information on scientific advances, but have helped to stimulate debate on the many ethical and moral issues which may arise. In view of the lay person's understandable interest in science with a direct personal impact, this growing awareness is often focused on the biomedical sciences. As a result, the health service is increasingly faced with patients who already have some idea of the options open to them derived from a variety of sources, and who expect to be actively involved in the choice of treatment for their condition. The old days of 'doctor knows best' are long gone, and the provision of 'good, relevant, unbiased and evidence-based information' should help patients to make informed decisions about the risks and benefits of the treatments they might be offered.

Although growing scientific awareness and associated demands for information are clearly major drivers of health service public information initiatives, there is an equally important need to challenge public ignorance. In this context, the provision of basic health information is designed to reduce inappropriate (and expensive) demands on the health service by making people aware of evidence about the self-limiting nature of common illnesses and the limitations of clinical intervention in such cases. At the same time, they need to be able to recognise potentially dangerous symptoms and present themselves for diagnosis and treatment at an early stage when intervention is likely to have the greatest (and least costly) chance of success.

Unfortunately many people, regardless of educational level, have only the haziest knowledge of how their bodies are constructed and work. A study carried out in 1995 compared the basic anatomical knowledge of 500 Asian and 500 non-Asian patients and found, for example, that only just over 60% of the non-Asians knew where their lungs were. Figures for the stomach, liver and other major organs were even lower although over 90% (but not 100%) could identify the location of the brain. In the widespread absence of basic knowledge of anatomy and physiology, and of the principles of first aid and home nursing, it is not surprising that some people are still demanding antibiotics for flu-like illnesses and calling out the doctor whenever a small child has a temperature, while ignoring persistent chest pain and unusual lumps. It is against this background that attempts to inform patients about the pros and cons of particular medical interventions need to be considered.

Empowerment through information provision is also a key part of the current government's emphasis on a consumerist agenda in the delivery of public services, given tangible form in initiatives like the Citizens' Charter, complaints procedures and the provision of performance monitoring information through league tables and other (often controversial) statistical indicators[159]. The avowed aim is to promote choice and improve the efficiency and effectiveness of services by giving users a measure of direct influence over providers. However, in the case of a health service which is now explicitly committed to priority-setting as a key element of management, the involvement of consumers is no doubt also seen as essential to ensure the legitimacy of decisions which may sometimes be seen as damaging to individual patient interests.

Critics argue that empowerment and choice are meaningless concepts in the context of a health service for which there is no realistic alternative for the bulk of users (except, perhaps, in the choice of GP), and that influence is limited in the absence of adequate funding or powers to respond to consumer concerns. There is certainly little evidence to show that consumer consultation or involvement initiatives have had any impact on issues that really matter to patients such as medical working practices and attitudes. As a result, the consumerist agenda may sometimes leave service users feeling both disempowered and resentful, while at the same time raising their expectations and their willingness to challenge authority. The dividing line between the informed patient and the difficult patient may be a thin one at times.

Empowerment or legitimation?

In the context of the individual doctor-patient relationship, the NHS clearly hopes that a public in possession of the evidence will make rational (i.e. NHS-preferred) choices about when it is appropriate to call on a doctor or other healthcare professional and, when under their care, which treatments are the most appropriate on the grounds of clinical effectiveness. Any attempt to reduce public ignorance about health and illness is to be welcomed, but it is perhaps unrealistic to expect that the mass of health service users can be educated into adopting the same attitudes and making the same choices as those who treat them. Despite the significant popular interest in science in the UK, most people are neither scientifically nor statistically literate and lack the basic skills to make truly informed judgements about complex medical or any other kind of scientific evidence.

The public reaction to last year's revelations about the potential risk factors of a range of oral contraceptives is a case in point. Though the government and its medical officers have been criticised for the way the issue was handled, considerable efforts were made to inform and reassure the public about the risk factors through the mass media. Despite this, many women panicked, stopped taking the pill and laid themselves open to the significantly greater risk of pregnancy and, in some cases, abortion. The understanding of relative risk remains at a low level, with major implications for the ability to judge evidence in many fields, and policy makers are actively looking at ways of improving matters[160]. Preventing irrational responses to scare stories about medical treatments, food safety and environmental hazards is clearly desirable, but could be extremely difficult given the widespread, and probably increasing, distrust of both political and scientific authority.

In the absence of advanced scientific and mathematical education for all, the provision of more information may do as much to confuse and frighten as to enlighten: a little knowledge is undoubtedly a dangerous thing. Moreover, it is rarely absolute, and some patients may well be disturbed by the growing realisation that scientific evidence is not always the truth that conventional wisdom holds it to be. How, for example, might a woman with breast cancer respond to the successive media reports of clinical trials on the use of Tamoxifen in recent months, some asserting that it should be given for no more than two years, others for five, still others for up to ten? The only certain knowledge she can derive from this information is that there is no firm evidence one way or the other, so far, on this particular form of therapy. As a result, she may be better informed about the realities of science but is unlikely to be either reassured or better equipped to make choices.

Thus we may be made more aware of the options available to us from the biomedical and other sciences but that does not necessarily mean we understand them any better or feel more confident about taking decisions. The public understanding of science movement seems, perhaps reluctantly, to have recognised this fact and now talks more frequently of public awareness – a much less rigorous term. There can be no true empowerment without real understanding, and NHS public information and involvement strategies may (like those of other public sector bodies) be little more than a useful tool for legitimising decisions already made by health care professionals and managers – a largely cosmetic device designed to promote the appearance of partnership.

Though this conclusion has a rather sinister quality about it, it is possible that the majority of patients would have little quarrel with such an approach. Although research studies have come to a variety of conclusions about how far people wish to participate, a significant proportion may be far more concerned with being informed than with taking decisions about their care, believing that this is still largely the responsibility of the professional. It is probable that few patients want genuine autonomy (one does, after all, go to a doctor specifically for expert advice) but they do want to be treated as responsible adults. Being informed about the options is a key element in establishing a relationship of mutual respect between doctor and patient, even if the latter's real freedom of choice is often circumscribed by ignorance.

The subjective element

However, even were it possible to educate all patients to the point where they could exercise genuine choice on the basis of medical evidence, it is by no means certain that all would make 'rational', scientifically determined decisions. Some will still deliberately opt for 'non-rational' choices which fly in the face of accepted evidence or which have no rigorously evaluated scientific basis. Users – patients and their families – cannot look on illness and its treatment in the same dispassionate way as providers. While patient and clinician have a great deal in common when considering how best to deal with illness, it is only the former who experiences direct suffering and, as the opening to Chapter 1 pointed out, the relief of suffering and the cure of disease are not necessarily the same thing. What the doctor sees as a rational, evidence-based approach to a particular condition, the patient may see quite differently.

'Non-rational' choices reflect the fact that the meaning of disease is personal and unique to the individual patient and cannot always be encompassed within the results of RCTs and other objective knowledge however scientifically well informed the patient may be. The fear of post-operative pain or side effects from drug treatment, a dislike of hospitals, a wish to avoid bodily interference whether from examination, surgery or drug treatment, a preference for 'natural' methods of healing and many other psychological factors can affect how the individual judges objective scientific evidence. So too can practical issues such as the impact of hospital admission or outpatient treatment on earnings, or on the ability to care for dependants. Any or all of these non-scientific considerations may make the patient reject the 'right' choice.

Many already do according to the Audit Commission's study of *What the doctor ordered*[101,Ch.4] which suggests that increasing demand from patients for services such as counselling and complementary therapies which have no rigorous evidence base is one reason behind GPs' currently low use of clinical effectiveness information in purchasing decisions. The booming demand for alternative 'natural' therapies of all kinds, perhaps fuelled by the vague environmentalism to which so many people subscribe, suggests that greater scientific awareness does not necessarily lead to the kind of decisions that the scientifically literate expect. In the wider context of science at large, it is interesting to note that those who buy *A brief history of time* or *The double helix* are just as likely to want books on astrology or the paranormal and, crucially, may view them as equally credible explanations of the world.

The combination of inadequate public knowledge, the personal and emotional dimension of disease, and patient empowerment may be a powerful force against the 'rational' process of EBM which better information may help to challenge but can never entirely eliminate. Its effects are already becoming apparent in one crucial area of EBM, the recruitment of patients to the randomised controlled trials which occupy such an important place in its development. Many doctors already have reservations about the impact of RCT recruitment on the traditional doctor-patient relationship, and there are genuine conflicts between individual and collective ethics in the medical field[161].

RCTs – and the gold standard information they provide – need a supply of willing volunteers but, as patients become more aware of (though not necessarily knowledgeable about) treatment options, and more willing to question medical authority, some are becoming resistant to the idea of denying themselves access to potentially effective treatments in the cause of scientific advance. Although patients are still often willing to accept expert advice as long as they feel they have been fully informed about all treatment options, they are increasingly resentful if their doctors refuse to give an opinion and offer a random trial instead[162]. Vital though the RCT may be to the collective good and to the development of evidence-based health care, it can look very much like a lottery to the individual patient.

Patient power is beginning to have a significant effect on the clinical trials world, where some Medical Research Council-backed trials are having to be abandoned because of patient resistance, especially to the random element of the RCT. Thus a trial to compare the effects of conventional chemotherapy and bone marrow transplant in the treatment of childhood leukaemia is in danger of collapse because parents demand the transplant which they believe to be more effective. Similarly prostate cancer trials are under threat because patients are insisting on surgery even though this may be much more dangerous than the cancer for older men. The cancer charity BACUP has tried to dispel the belief that clinical trials are about 'wild and high-risk experimentation', but many patients are clearly not convinced[163].

Some commentators argue that this problem can be overcome by information/education strategies and – crucially – by the willingness of doctors to admit the uncertainty of much of their knowledge (an argument which has also been put forward in the wider context of the public understanding of science). Involving patients in the planning stage of clinical trials may also help. However, the degree to which these strategies will change attitudes must be open to doubt. Making people aware that science is not the unambiguous truth they have traditionally believed it to be may encourage more rather than less subjective decision making. If 'rational' science cannot provide certainty and safety, people will look elsewhere for help in making difficult choices.

Nor is a high degree of knowledge about the nature of scientific inquiry necessarily a guarantee of support for the clinical trial process. AIDS patients are probably among the best informed of all and are only too well aware that 'their survival interests and the interests of researchers trying to generate knowledge do not necessarily coincide'[164]. Non-compliance and drop-out rates in AIDS research can reach as high as 80% as the 'lab rats' revolt against a research community which they no longer trust. Many are turning to unproven alternative therapies in the belief that 'the establishment lies anyway' and that 'drugs are the cause of AIDS'.

Respect for the patient's view

Although the empowerment of patients through information provision may not necessarily have the outcomes the NHS expects, it is a strategy the health service clearly intends to pursue and one in which EBM will play a major part. Evidence-based information does not simply meet the patient's legitimate desire for greater knowledge about treatment options and outcomes, it can also be used to explain 'difficult choices' within a resource-constrained health service. Increasing numbers of patients may suspect that medical decision making within the internal market is being driven primarily by budgetary constraints, and EBM information may help to counter this belief. However, any attempt to use such information to disguise or justify outright rationing is likely to rebound on the NHS.

The patient perspective of EBM is not simply concerned with redressing the traditional imbalance in knowledge and power between patient and doctor. Though it is clearly desirable to provide health service users with more information about treatment options and outcomes, if they want it, there is a parallel need to improve the flow of patient-centred information into the generation of evidence itself[165]. Suffering, as Archie Cochrane recognised, is not just about disease, and a health service dedicated to the relief of suffering needs to move beyond the purely medical model to encompass the complex psychological and social perspectives of those it serves. Primary medical research and the assessment of health care outcomes need to be informed by these perspectives if they are to deliver the greatest benefit. As a result, patients' organisations are becoming increasingly vocal in their demands for involvement in setting research directions and promoting the transfer of good quality evidence into practice as ways of improving the quality of life of those they represent.

Consumer involvement, including the provision of information, is a guiding priciple of the Cochrane Collaboration, and a paper by Hilda Bastian of the Australasian Cochrane Centre looks in some detail at the issues it raises. She notes that the evidence-based approach to health care may lead to effective new (or existing) treatments being given greater emphasis, to the benefit of patients, but it also leads to pressure for less effective ones to be abandoned[166]. However, decisions on what is, or is not, effective are rarely clear cut and are at least in part value judgements. For this reason, they should not be made solely by professionals and service providers. Health services are there to care for people, and it is vital to give them a voice to counterbalance the huge power exercised by researchers, health care professionals and managers in setting the NHS research and practice agenda.

'Specialists' views of the world are by definition unbalanced and incomplete' and are not always consistent with the public interest, according to Hilda Bastian. She uses the Cochrane Pregnancy and Childbirth Database as an example, noting that 'issues of obvious significance to women are often conspicuous by their absence' from the research it contains. Many projects with impeccable scientific credentials report the impact on neonatal outcomes of procedures performed on women, but there are almost none concerned with the impact of pregnancy, childbirth and associated medical interventions on the quality of life of women themselves. Thus there is virtually nothing on the relief of pain (backache, for example) during pregnancy, little on post-delivery pain, and none on the long term consequences of pain relief during labour. Equally, there is nothing on pain after caesarean section, despite the frequency with which the procedure is carried out, or on the relief of pain for breastfeeding mothers.

Accepting the need for greater consumer involvement to ensure a more balanced research and practice agenda is one thing; persuading people of the value of participation, and ensuring that it works effectively is another. Hilda Bastian looks at some of the problems involved in preventing the marginalisation of lay people in settings dominated by 'experts', commenting that even the Cochrane Collaboration is not immune from the charge of insider elitism. This issue has to be addressed by the Collaboration in the development of its Consumer Network, and by health (and other public) services more widely, if consumer involvement is to deliver the expected benefits. The views of lay people will inevitably be different from those of technical experts, but it is in these differences that their value lies. Unless they are respected, and acted upon where appropriate, involvement and consultation are no more than a sham and will be rejected as such[167].

The Collaboration's Consumer Network is being pioneered by the Australasian Cochrane Centre with the basic aims of fostering 'an environment within the Collaboration which encourages and facilitates consumer input' and promoting 'consumer groups' participation in, and use of, systematic reviews of the effects of health care'. Its roles and activities include:

- Identifying people with an interest, primarily by canvassing appropriate consumer organisations.
- Producing and distributing information to encourage greater awareness of, and participation in, the Cochrane Collaboration by consumer groups with an interest in health.
- Fostering, in conjunction with the individual Cochrane Centres, liaison between consumer representatives in the Collaboration.

- Encouraging the development of critical appraisal and database skills among consumers.
- Conducting, coordinating and encouraging reviews and research relating to consumer interaction with health care systems and their participation in research.

NHS consumer involvement initiatives

The NHS has responded to the demand (and need) for greater individual and collective involvement by consumers in a variety of ways which are usefully summarised by the King's Fund in *Health care partnerships*[168]. They include the Patients' Charter published in 1991 which, for example, gives the right 'to have any proposed treatment, including any risks involved in the treatment and any alternatives, clearly explained to you before you decide whether to agree to it'[169]. Information provision, as discussed below, is clearly important in this context but so too are changes in professional attitudes and practices which need to be inculcated and reinforced by developments in medical education to promote better inter-personal communication skills and challenge the tradition of benevolent paternalism that has characterised the doctor-patient relationship[170].

Initiatives to involve patients in the planning and delivery of services, provide performance information and ensure redress in the event of problems include the Patient Partnership strategy which has developed from the Patient Empowerment Focus Group set up in 1993[171]; the *Local voices* initiative[172]; the publication of league tables and the results of patient satisfaction surveys[173]; and the introduction of complaints procedures. On the research side, the Central Research and Development Committee has a new standing advisory group on consumer involvement in the R&D strategy[174], while the UK Clearing House on Health Outcomes has examined how the patient and carer perspective can be incorporated within the outcome assessment process.

The College of Health (on behalf of the NHS Clinical Outcomes Group) has also examined the issue of consumer involvement in clinical audit following considerable criticism of earlier NHS insistence that audit should be solely a mechanism for professional peer review[175]. However, despite Department of Health instructions that providers should 'develop mechanisms to ensure successful patient/carer input to clinical audit processes', many doctors serving on audit groups still appear hostile to lay involvement. Setting up new mechanisms for involvement in audit or any other 'expert' activity is rarely enough on its own; there must also be active attempts to support lay participants with training and information, and to challenge marginalisation.

Consumer health information services

Outside the boundaries of the NHS, the King's Fund is a key focus for research on increasing patient involvement in health care and health services through its Promoting Patient Choice programme of studies which includes research into the role of the consumer health information sector in the dissemination of treatment outcomes to the public[176]. Much of this has been carried out by the Help for Health Trust, a national charity whose mission is 'to help people become active partners in their own health care by providing them with information'. As well as delivering health information services to the public within a national network of centres, it is also active in consumer health information research and consultancy, and provides information service support to the Wessex Institute of Public Health Medicine.

Initial survey evidence from the Trust suggested that 42% of consumer health information services reported 'frequently' receiving requests to support treatment decisions, some for reassurance after treatment had started but 27% during the early stages before the treatment decision[177]. However, despite the fact that the number and variety of consumer health information services have increased substantially in recent years, evidence from a later study showed a significant unmet consumer need for information on tests, treatments and the management of conditions, and low awareness of the information services available[178].

Five consumer information service pilot projects have now been evaluated and have focused attention on the need for information provider training in all stages of the dissemination process including eliciting the level and type of information required, searching for relevant information, selecting appropriate material, and helping clients to interpret it where necessary[179]. The issue of how far information providers should go in interpretation and education is a delicate one given the expressed aim of such projects to empower clients, but adult education techniques may well be of value. Improvements in the quality of health information materials are also needed, not only to ensure that they are based on the best available evidence (few currently are) but also to provide real choices. In *Health care partnerships*[168] the King's Fund identifies the following characteristics of good quality patient information which should:

- 'inform the patient about the clinical condition and include information about all available treatment or management options, including non-intervention;
- provide comprehensive and unbiased information about outcomes (risks and benefits) based on systematic review of the research evidence;
- outline uncertainties and gaps in scientific knowledge;
- be simple in language and design, attractive and easily understood;
- cater for a variety of users including people who do not speak English and people with sensory or learning disabilities;
- be regularly reviewed and updated;
- be integrated into a planned programme for shared clinical decision-making;
- involve users and professionals in the development and evaluation of the materials.'

Written forms of information provision – including the ubiquitous patient leaflet – are likely to remain the most common medium for the time being[180], and various initiatives are underway to produce materials which are firmly evidence-based. The NHS Centre for Reviews and Dissemination has cooperated with Bedfordshire Health Authority to adapt its *Effective Health Care Bulletin* on glue ear for a lay readership, and with Buckinghamshire Health Authority and the Cataract Focus Group to produce a leaflet for cataract patients. It is also collaborating with MIDIRS (Midwives Information and Resource Service) in the Informed Choice in Maternity Care Initiative. This aims to produce a parallel range of evidence-based leaflets for professionals and childbearing women on subjects of both general relevance (such as ultrasound scanning) and of more limited interest (such as the management of breech presentation)[181].

The digital future

There is also increasing interest in the use of advanced information and communication technologies to disseminate health information to the public. The King's Fund is evaluating a number of American interactive videos, and developing its own multi-media packages involving combinations of videos, CDs, leaflets and computer-assisted learning packages. Other video trials include a project at University College London Medical School which is developing and testing videos on hysterectomy and prostate cancer.

Interactive videos may be a particularly promising medium given the fact that many people rely on visual media for information and often absorb and retain knowledge more readily from these sources than from conventional leaflets and books. They may also promote more genuinely shared decision making between patient and doctor. However, they are expensive – £4,700 for the specialised machinery needed to run the UCL videos, plus a further £1,000 for each of the laser disks that control the programme in line with the specific characteristics and needs of individual patients.

In the future, people will doubtless be able to gain access to health information from the Internet either at home or via public access kiosks[182], although Bob Gann of the Help for Health Trust warns against over-optimism in this area. Despite its rapid growth, the Internet is still largely the preserve of young, high income men and it remains a baffling, even frightening, concept to many and well beyond the financial resources of a significant proportion of the population. Moreover, the message is far more important than the medium, and the anarchic nature of the Net could raise major problems in ensuring the quality and reliability of what is provided to consumers. Similar difficulties may be experienced with the private sector development of standard, non-interactive videos for patients unless rigorous (and transparent) quality assurance procedures are followed[183].

The Centre for Health Information Quality

The importance of quality assurance is reflected in the title of the most recent initiative in the consumer health information field. The Centre for Health Information Quality, established in April 1997 at the Help for Health Trust, follows commitments in the White Papers on *The National Health Service: a service with ambitions*[46, Ch.2] and *Primary care: delivering the future* (Stationery Office, Dec 1996. 60pp. Cm 3512). Its objective is 'to act as a source of expertise and knowledge for the NHS and patient representative groups on all aspects of patient information with the aim of improving the NHS's capability, competence and capacity to provide good, evidence based patient information' (Cm 3512, para. 4.19).

Though based at the Help for Health Trust, the Centre is a partnership that also involves the Wessex Institute for Health Research and Development (providing health promotion expertise and links with the Health Technology Assessment Programme), the Critical Appraisal Skills Programme (providing training in critical appraisal skills and evidence-based health), and Buckinghamshire Health Authority (providing practical local experience of user involvement. Funded for three years by the NHS Executive, it is expected to become self-financing after March 2000.

The Centre will act as a clearing house and centre of excellence on all aspects of patient information, and its work will include the development of quality tools to evaluate information for patients; a demonstration collection of high quality information materials in print and non-print formats; a database of good practice in the creation of high quality patient information (to be made available via the World Wide Web); a skills clearing house; a programme of training for information producers; and a network of topic-based patient panels to review available material in the light of the real needs and priorities of patients. Strong links with the NHS R&D programme through the National Coordinating Centre for Health Technology Assessment (based in the same building) will also be a feature of the Centre's work.

This new initiative should help to ensure that the growing enthusiasm for informing and involving patients in their health care is tempered by an awareness of the importance of quality assurance in this sensitive area. Despite the frequently voiced demand of patients' representatives to 'give us the information and let us decide', most of us still have a limited ability to judge the value of scientific or technical information, especially in those cases where it is equivocal. Without proper quality control to ensure that information is firmly based on the best evidence and presented in accessible ways, the Patients Association's call for 'virtual reality health councils' giving access to NHS performance measurement data and other health information may not, as its chairman has suggested, give patients 'power over themselves, their illnesses and their lives' (as reported in the *Independent*, 11 May 1996 p6). Rather, it may merely serve to confuse and worry, possibly raising unrealistic expectations and certainly doing little to advance the cause of rationality in health care provision which lies at the heart of the EBM movement.

(158)
CARING for people: community care in the next decade and beyond
Department of Health
HMSO (now Stationery Office), 1989. 106pp (Cm 849)
Creates a framework for active participation by health service users and carers in the design of care plans, and establishes mechanisms for collective involvement in community care planning.

(159)
CONSUMERISM and the NHS: an annotated bibliography
Buckland, S and others (editors)
Social Services Research and Information Unit: Portsmouth, and Portsmouth University School of Social and Historical Studies, Jul 1993. 81pp (Joint SSRIU/HIRS Occasional Paper 1)

(160)
THE SCIENCE of perceiving risk
Taylor, I
Chemistry and Industry, 2 Dec 1996 (23) p956
The then Minister of Science and Technology writes briefly on the need to improve the assessment of risk and the effective communication of risk information to the public. The aim is to reduce anxiety and irrational behaviour sparked off by scare stories in the press.

(161)
CLINICAL biostatistics: an introduction to evidence-based medicine
Dunn, G; Everitt, B
Edward Arnold: London, 1995. 154pp
An undergraduate text book which emphasises the key importance of statistical skills in modern evidence-based medicine, and gives a useful introduction to the nature of RCTs and meta-analyses, and the ethical and practical problems that may be associated with them. For more on the ethical aspects of clinical trials and EBM
See also: The ethics of randomised controlled trials: a matter of statistical belief? by J L Hutton (Health Care Analysis, May 1996 4(2) pp95–102)
Evidence-based medicine and ethics, by T Hope (Journal of Medical Ethics, Oct 1995 21(5) pp259-60)

(162)

MEASURING patients' desire for autonomy: decision making and information-seeking preferences among medical patients

Ende, J and others

Journal of General Internal Medicine, Jan/Feb 1989 4(1) pp23-30 (+ comment in Journal of General Internal Medicine, Jul/Aug 1989 4(4) pp360-61)

Develops and tests an instrument for measuring patients' preferences for two dimensions of autonomy: the desire to make medical decisions, and the desire to be informed. Finds that patients prefer decisions to be made principally by clinicians, especially when they are severely ill, but also very much wish to be informed.

(163)

UNDERSTANDING clinical trials

BACUP: British Association of Cancer United Patients, 3 Bath Place, Rivington Street, London EC2A 3JR, 1996. 24pp

Attempts to allay public fears about participating in clinical trials, noting that there are good personal as well as altruistic reasons for involvement. Patients in clinical trials may have a better survival rate than others, regardless of the efficacy of the treatment under consideration, because their progress is more closely monitored by more people.

(164)

AIDS and the lab rats

Schüklenk, U

Science and Public Affairs, Winter 1996 pp54-57

Looks at the issues surrounding the declining confidence of many people with AIDS in the clinical trials process, and in the motivations of those who carry them out. Many believe they are regarded simply as experimental animals, and that research programmes benefit the scientists (often in monetary terms) more than themselves. Looks at ways of overcoming a mistrust which can lead patients to deny themselves access to potentially beneficial drugs simply because they emanate from the medical establishment. For more on the impact of AIDS activism, and on the more general implications of patient power for clinical trials

See also: Impure science: AIDS, activism and the politics of knowledge, by S Epstein (University of California Press, 1996. 466pp)

Patient preferences in randomised trials: threat or opportunity?, by D J Torgerson and others (Journal of Health Services Research and Policy, Oct 1996 1(4) pp194-97)

(165)

ASSEMBLING the evidence: patient-focused outcomes research

Coulter, A

Health Libraries Review, Dec 1994 11(4) pp263-68, 8 references

Argues the case for much greater patient involvement in outcomes research using as an example the risks and benefits of hysterectomy where research into quality of life issues that matter to women has been notable by its absence. For more on patient involvement in outcomes research including the proceedings of a conference

See also: Consumer involvement in outcomes measurement: what are the barriers? by R Wiles (Critical Public Health, 1993 4(4) pp2-3)

But will it work doctor? Involving users of health services in outcomes research, edited by M Dunning and G Needham (Consumer Health Information Consortium, 1994. 52pp)

(166)

THE POWER of sharing knowledge: consumer participation in the Cochrane Collaboration

Bastian, H

Cochrane Collaboration Paper, Dec 1994, 41 references (Available via the Cochrane Collaboration home page at **http://hiru.mcmaster.ca/cochrane/default.htm**)
A two part analysis focusing first on differing concepts of the public/consumer role, the role of consumer participation, the representation of consumer perspectives, and the debate over terminology (should we be called consumers, patients, sufferers, people etc., etc?). The second part deals with the Cochrane Collaboration's attempts to make itself more accessible to consumers.

(167)

THE LISTENING blank

Donaldson, L

Health Service Journal, 21 Sep 1995 105(5471) pp22-24
Outlines NHS attempts to involve the public in health care planning, and indicates what makes for successful consumer involvement. For further comment on public consultation exercises, some of which are perceived by the public as a sham
See also: Local voices in an internal market: the case of community health services, by S Pickard and others (Social Policy and Administration, Jun 1995 29(2) pp135-49, 22 references)
Does public opinion matter? by M Rigge (Health Service Journal, 7 Sep 1995 105(5469) pp26-27, 3 references)

(168)

HEALTH care partnerships: debates and strategies for increasing patient involvement in health care and health services

Farrell, C

King's Fund, 11-13 Cavendish Square, London W1M 0AN, 1996. 41pp, bibliography pp36-41 (Promoting Patient Choice 1)
Looks at the conceptual confusion over what is meant by involvement, participation, empowerment, patient, user and consumer; discusses collective and individual lay involvement in the NHS; and makes recommendations for change in respect of information provision to patients, the training of health care professionals, the establishment of a health service users' council, the strengthening of Community Health Councils, increased accountability in primary care, and greater openness in the NHS system. For more from the King's Fund on improving patient choice through information provision
See also: Evidence-based patient choice, by T Hope (King's Fund, 1996. 39pp)

(169)

THE PATIENT'S charter: raising the standard

Department of Health, Richmond House, 79 Whitehall, London SW1A 2NS, Nov 1991. 23pp
For a response to the Charter by the Consumers' Association, which argues that it will not fulfil its objectives without the widespread adoption of consumer appraisal techniques in the NHS, and for a briefing from the Association of Community Health Councils
See also: Patients as experts: consumer appraisal of health services, by K Steele (Public Money and Management, Oct/Dec 1992 12(4) pp31-37, 11 references)
The Patient's Charter: the patient's perspective (Association of Community Health

Councils for England and Wales, Earlsmead House, 30 Drayton Park, London N5 1PB, Sep 1994. 19pp)

(170)
COMMUNICATION skills
Gabbay, J and others (editors)
Royal College of Physicians, Faculty of Public Health Medicine, 11 St Andrews Place, London NW1 4LE, Nov 1995. 100pp
For a study casting light on the unsatisfactory nature of many doctor-patient relationships, and for the General Medical Council's guidance on good practice in this and other areas of medical practice
See also: Whose standards: consumer and professional standards in health care, by C Williamson (Open University Press: Buckingham, 1992. 155pp)
Duties of a doctor: guidance from the General Medical Council (General Medical Council, 44 Hallam Street, London W1N 6AE, 1995. 56pp)

(171)
PATIENT partnership: building a collaborative strategy
NHS Executive, Anglia and Oxford Region, 6-12 Capital Drive, Linford Wood, Milton Keynes, Buckinghamshire MK14 6QP, Jun 1995. 17pp

(172)
LOCAL voices: the views of local people in purchasing for health
NHS Executive, Richmond House, 79 Whitehall, London SW1A 2NS, Jan 1992. 24pp
For more comment, from a variety of perspectives, on the development of links with local people
See also: Involving the community: guidelines for health service managers (National Consumer Council, 20 Grosvenor Gardens, London SW1W 0DH, 1992. 11pp)
Public participation in health care: involving the public in health care decision making: a critical review of the issues, by B H Gurney (University of Cambridge, Health Services Research Group, 1994. 99pp)
Listening to local voices, edited by C Heginbotham (National Association of Health Authorities and Trusts (now NHS Confederation), Birmingham Research Park, Vincent Drive, Birmingham B15 2SQ, 1993. 38pp. Research Paper 9)
Involving local people: examples of good practice (NHS Executive, 1994. 29pp)

(173)
THE MEASUREMENT of patient satisfaction
Carr-Hill, R
Journal of Public Health Medicine, Sep 1992 14(3) pp236-49
A comprehensive analysis of the various perspectives and definitions of satisfaction, and the difficulties of carrying out patient satisfaction surveys. Many surveys show very high levels of satisfaction (possibly associated with feelings of gratitude or vulnerability while ill), and have been widely criticised for conceptual weakness and managerial bias.

(174)
CONSUMERS and research in the NHS: consumer issues within the NHS
NHS Executive, R&D Directorate, Quarry House, Quarry Hill, Leeds LS2 7UE, Aug 1995. 3 booklets
The booklets comprise a foreword from Sir Michael Peckham; a useful analysis of consumer issues and related research by Professor Mildred Blaxter of the University of East

Anglia (including a bibliography); and the report of a workshop on involving consumers in
local health care. Professor Blaxter's review includes comment on involving consumers in
the research agenda, but covers many other aspects of the consumer debate including the
concept of patient satisfaction; consumer priorities; the differences between consumer and
professional perspectives; the measurement of needs and outcomes; and consumer
consultation.

(175)

**CONSUMER involvement initiatives in clinical audit and outcomes: a review of
developments and issues in the identification of good practice**
Kelson, M
College of Health, St Margaret's House, 21 Old Ford Road, London E2 9PL, and
Department of Health Clinical Outcomes Group, Feb 1995. 62pp, 298 references
Catalogues and reviews the published literature on both individual and collective
consumer involvement in the health service, with particular reference to clinical audit.
Finds confusion over terminology and methodology which hamper the search for best
practice models, and shows little evidence of systematic inclusion of consumers in audit/
outcomes measurement despite the calls of the Department of Health and NHS Executive.
For more on involvement in audit, including the College of Health's own guidelines on
qualitative consumer audit to complement scientific clinical audit
See also: User involvement in medical audit: a spoke in the wheel or a link in the chain,
by N Joule (Greater London Association of Community Health Councils, 356 Holloway
Road, London N7 6PA, Sep 1992. 54pp)
Consumer audit guidelines (College of Health, 1994. 147pp)

(176)

CONSUMER health information: the growth of an information specialism
Gann, B
Journal of Documentation, 1991 47(3) pp284–308
Looks at the growth of consumer health information services in the UK and abroad,
noting a wide variety of settings including telephone helplines, health information shops,
public and medical libraries, community health councils and health promotion
departments.

(177)

**DISSEMINATION of information on treatment outcomes by consumer health
information services: report to King's Fund**
Buckland, S; Gann, B
Help for Health Trust, Highcroft, Romsey Road, Winchester, Hampshire SO22 5DH,
Nov 1994. 75pp (Tel: 01962 849100; email: admin@hfht.demon.co.uk; URL: **http://
www.hfht.demon.co.uk/**)
Presents initial survey evidence on the types of demand experienced by 53 consumer
health information services including national voluntary organisations, regional health
information services and members of the Consumer Health Information Consortium. For
additional reports from the Office of Health Economics on the sources of health
information used by the public and their effectiveness in changing attitudes and behaviour
See also: Health information and the consumer (Office of Health Economics, 12
Whitehall, London SW1A 2DY, May 1994. 12pp. Briefing 30)
Health information and the consumer, edited by J Griffin (Office of Health Economics, 12
Whitehall, London SW1A 2DY, 1995. 77pp. [Papers from a symposium held in
November 1994])

(178)

UNMET need for health information: report to Nuffield Provincial Hospitals Trust

Buckland, S

Help for Health Trust, Highcroft, Romsey Road, Winchester, Hampshire SO22 5DH, Jan 1995. 190pp, bibliography

Presents the results of an exploratory survey of the extent and nature of unmet need for consumer health information, based on in-depth interviews with 40 members of the public. They show that information from nursing and medical staff was often limited, especially in respect of information about conditions and illnesses. Family, friends and books are the most commonly used sources of information prior to consultation with health professionals, and there is low awareness and use of self-help groups or telephone helplines as sources of information. Also includes a survey of the literature on unmet health information needs from 1987 onwards. For a summary

See also: Unmet need for health information, by S Buckland (Health Libraries Review, Jun 1994 11(2) pp82-95)

(179)

DISSEMINATING outcomes information to consumers: evaluation of five pilot projects: report by the Help for Health Trust to the King's Fund on Phase Two of the project

Buckland, S; Gann, B

Help for Health Trust, Highcroft, Romsey Road, Winchester, Hampshire SO22 5DH, 1996.

(180)

WRITING leaflets for patients: guidelines for producing written information

Secker, J; Pollard, R

Health Education Board for Scotland: Edinburgh, 1995. 68pp

(181)

A PILOT study of 'informed choice' leaflets on positions in labour and routine ultrasound

NHS Centre for Reviews and Dissemination, University of York, Hesington, York YO1 5DD, and others, Dec 1996. 94pp (CRD 7)

The study was carried out in association with the Midwives Information and Resource Service and the University of London Social Science Research Unit. For another recent evaluation of the role of leaflets in promoting informed choice about ultrasonography by women in early pregnancy

See also: Informed choice for users of health services: views on ultrasonography leaflets of women in early pregnancy, midwives, and ultrasonographers, by S Oliver and others (British Medical Journal, 16 Nov 1996 313(7067) pp1251-53)

(182)

INFORMATION on the Internet: consumer guides to health care

Lindsay, B

British Journal of Health Care Management, 1996 2(8) pp445-47

Highlights the self-help potential of the new technology. For more on the possibilities offered by interactive media

See also: Interacting: multimedia and health, by J Leonard (Health Education Authority, Hamilton House, Mabledon Place, London WC1H 9TX, 1994. 126pp)
The future's bright: the future's digital: information technology and health information towards the millennium and beyond, by S Wallace (King's Fund, 11-13 Cavendish Square, London W1M 0AN, 1996. 80pp. Promoting Patient Choice 5)

(183)
VIDEOS for patients: what you really need to know
Videos for Patients Limited, 122 Holland Park Avenue, London W11 4UA, 1996. 8pp
An advertising brochure from a company set up by John Cleese in association with Dr Rob Buckman. It has produced 45 videos, price £15.00 each to the patient, on a wide range of common conditions of varying degrees of seriousness. Attempts are made to ensure quality: each video is 'checked by two independent and established authorities in the field' prior to production, with one of the authorities supervising recording. However, it is not clear how these authorities are selected and nor are they named.

INDEXES

1. Authors

Numbers relate to references

2. Corporate bodies

Numbers relate to references

3. Subject index

Numbers relate to pages

ABOUT THE BRITISH LIBRARY

The Science Reference and Information Service (SRIS) is a world leader for information in science, technology, medicine, business, patents and the social sciences. As part of the British Library, it provides a unique resource for the UK science, business, government and academic communities by offering access to a wide range of literature and the largest patent collection anywhere in the world.

The specialist information services in business, science and technology, the environment, health care, patents and social policy are able to give expert help to users. They offer free quick enquiry services for simple queries and competitively priced research services for more complex enquiries.

Users can do their own searching by subscribing to SRIS' online services, BLAISE-LINK, which offers subscribers access to the biomedical and toxicological files of the US National Library of Medicine, and IRS-Dialtech, the UK national centre for the European Information Network Service's host computer.

More than 170,000 people visit SRIS' four central London reading rooms each year and thousands more use the specialist information services. Whether they need access to online databases, specialist information, translations, access to our research collections or training in making the most of information, SRIS can help. Although our users come from all walks of life they have in common a need for high quality information and the sort of expert help SRIS' professional staff are able to provide.

To complement the services of SRIS, documents can be supplied to customers offsite via the Library's Document Supply Centre (DSC). The Centre provides the most comprehensive photocopy/loan service in the world and has unrivalled collections of journals, books, conference proceedings and theses, as well as the UK National Reports Collection. For more information on DSC services telephone 01937 546060 or fax 01937 546333, email dsc-customer-services@bl.uk.

To find out more about our collections and specialist services contact:

SRIS Marketing and Public Relations
25 Southampton Buildings
London WC2A 1AW
Tel: 0171 412 7959
Fax: 0171 412 7947
Email: sris-customer-services@bl.uk.

Anyone with an Internet connection can find out more about SRIS on http://www.bl.uk/sris/ or you can access our catalogue on http://opac97.bl.uk.

Guide to Libraries and Information Sources in Medicine and Health Care

Looking for information on a specific subject in medicine or health care? *Guide to Libraries...* covers hundreds of subjects - from anorexia to homoeopathy, meningitis to women's health - helping you pinpoint the right organisation for all your information needs.

The *Guide* provides detailed information on 718 organisations, including: hospital libraries, professional associations, company libraries, charities, voluntary bodies and support groups. All of the organisations listed are willing to accept 'outside' enquiries - saving you hours of research time, and dozens of frustrating phone calls, in looking for specialist information.

The British Library, February 1997
192 pages approx., 297x210mm, paperback
ISBN 0-7123-0839-3
Price £39 (UK postage included, overseas postage extra)

For further information contact Tony Antoniou at the Science Reference and Information Service, 25 Southampton Buildings, London WC2A 1AW. Tel: 0171-412 7471. Publications are on sale to personal callers at SRIS.

ORDER FORM

Orders should be sent to:

Turpin Distribution Services Ltd, Blackhorse Road, Letchworth, Herts SG6 1HN. Telephone: 01462 672555, Fax: 01462 480947, Email: turpin@rsc.org

Callers should ask for 'The British Library Section'.

Prices quoted include UK postage and packing. Extra postage is charged for all overseas orders. Contact Turpin for details.

To pay by credit card:

Payment by credit card is accepted for ACCESS, VISA, MASTERCARD and EUROCARD. Telephone 01462 672555 to place your order.

To pay by cheque or proforma invoice:

Fill in the order form opposite.
Cheques should be made payable to 'The British Library'

☐ I enclose a cheque made payable to 'The British Library' for £

☐ Please invoice me

Please send me.................copies of *Guide to libraries and information sources in medicine and health care*, 2nd ed, ISBN 0-7123-0839-3, price £39

Name_____

Position _____

Organisation _____

Address _____

_____

Invoice address if different from above _____

_____

Can scientists always be trusted to tell the truth?
Scientific Deception will help you decide.

Accusations of fraud, deception and malpractice have become increasingly common as the media and the public have grown more and more sceptical about scientists and about the research process.

Scientific Deception, written by the respected writer and bibliographer Lesley Grayson (editor of the journal *Science, Technology and Innovation*), the guide looks at the entire range of possible misconduct from the most blatant fraud to relatively innocent self-deception. The author explores the reasons for such behaviours before going on to discuss responses and policy implications, drawing on material from the United States, Australia and Europe, including the UK. Each chapter includes annotated lists of references, including the most up-to-date (1995) material.

Until now there has been little discussion in the UK scientific community about how to stamp out slipshod or even dishonest work and to ensure that high standards are maintained. With pressure growing for more concerted action (especially in the field of medicine), and for the setting up of an independent investigatory body, *Scientific Deception* is essential reading.

Scientific Deception: An Overview and Guide to the Literature of Misconduct and Fraud in Scientific Research.
The British Library
November 1995. 120 pages approx.
297x210mm, paperback
ISBN 0-7123-0831-8
Price £24.50 (UK postage included, overseas postage extra)

For further information contact Tony Antoniou at the Science Reference and Information Service, 25 Southampton Buildings, London WC2A IAW. Tel: 0171-412 7471. Publications are on sale to personal callers at SRIS.

The world's leading resource for scholarship, research and innovation